ACING SCIENCE

FEMINIST TECHNOSCIENCES
Rebecca Herzig and Banu Subramaniam, Series Editors

ACING
SCIENCE

COMPULSORY SEXUALITY AND
ASEXUAL POSSIBILITIES

KRISTINA GUPTA

UNIVERSITY OF WASHINGTON PRESS
Seattle

Design by Katrina Noble   Composed in Calluna

UNIVERSITY OF WASHINGTON PRESS
uwapress.uw.edu

Cataloging information is available from the Library of Congress

LIBRARY OF CONGRESS CONTROL NUMBER 2025029048
ISBN 9780295754253 (hardcover)
ISBN 9780295754260 (paperback)
ISBN 9780295754277 (ebook)

For EU product safety concerns please contact Easy Access System Europe Oü 16879218, Mustamäe tee 50, 10621, Tallinn, Estonia, gpsr.requests@easproject.com, +372 56 98939.

♾ This paper meets the requirements of ANSI/NISO Z39.48-1992 (Permanence of Paper).

To Gabriel Castro-Gupta

CONTENTS

Acknowledgments   ix

INTRODUCTION: Hegemonic Science, Compulsory Sexuality, and Asexual Possibilities   1

CHAPTER 1.   "There Is a Great Deal of Denial in This Population": Euro-American Medicine and Compulsory Sexuality   21

CHAPTER 2.   Sex in the Machine: The Neuroimaging of Sexual Desire   48

CHAPTER 3.   Pandas, Voles, and Rams: Asexual Phenomena in Nonhuman Animals   75

CHAPTER 4.   Amoebas Are Us? Asexual Reproduction and the Category of Sex   103

CONCLUSION: Toward an Antinaturalist, Intersectional Ace Politics   131

Notes   149

References   181

Index   229

ACKNOWLEDGMENTS

IT TAKES A COMMUNITY TO WRITE A BOOK. THE ONE BEHIND THIS BOOK is particularly large, as I began work on this project as a PhD student in the Department of Women's, Gender, and Sexuality Studies at Emory and have continued to work on it over the past fifteen years. I am deeply grateful to my mentors at Emory, especially my doctoral advisor, Rosemarie Garland-Thomson, and my doctoral committee members, Sander Gilman and Mark Risjord. I am also grateful to the faculty and students who were involved with the Emory Neuroethics Program (especially Karen Rommelfanger and Gillian Hue) and the Emory Center for Mind, Brain, and Culture (especially Laura Namy and Robert McCauley). I could not have survived without the intellectual community and friendship offered by my fellow graduate students, especially Cyd Cipolla, Angela Willey, David Rubin, Julia Haas, Anson Koch-Rein, Nikki Karalekas, Kira Walsh, Megan Friddle, and Harold Braswell.

I am indebted to my friends and colleagues at Wake Forest University, who have supported me both professionally and personally, especially Mir Yarfitz, Stephanie Koscak, Tanisha Ramachandran, Amanda Gengler, Simone Caron, Michaelle Browers, Barry Trachtenberg, Steve Boyd, Melissa Jenkins, Rian Bowie, T. H. M. Gellar-Goad, Sarah Raynor, Elizabeth Clendenning, and Alessandra Von Burg. I am grateful for the community provided by the Wake Forest Disability Studies Initiative and for my codirector, Adam Kadlac, and for the activist community provided by Wake Forward. Special thanks to my students for their ongoing engagement and enthusiasm.

I am fortunate to work within a wonderful community of scholars in the field of asexuality studies, especially K. J. Cerankowski, Ela Przybylo, Ianna Hawkins Owen, and Megan Milks. I remain deeply appreciative of the ace folks who agreed to be interviewed in 2010–11, as their stories continue to serve as inspiration for my work. I am thankful for the ongoing work of aspec activists and community builders, who make the world a better place for everyone.

Many friends and colleagues provided direct feedback on the manuscript. My department colleagues did so for an initial draft of the introduction. Chapter 1 is a heavily revised version of a chapter from my dissertation, which was informed by the commentary of my committee. Mir Yarfitz provided feedback on a more recent version of chapter 1. Chapter 2 includes a revised version of a post on the *Neuroethics Blog*, which was reviewed by Karen Rommelfanger. This chapter was also informed by my participation in the Neurogenderings Network and collaborations with Cyd Cipolla as Neuroethics Fellows. I also received input on this chapter from attendees at the National Women's Studies Association (NWSA) 2022 conference. The seeds of chapter 3 were written for a class I took at Emory with Kim Wallen, who provided generous feedback. I also received suggestions for this initial work from attendees at the 2012 Association for Feminist Epistemologies, Methodologies, Metaphysics and Science Studies (FEMMSS) conference. Anson Koch-Rein reviewed a more recent version of chapter 3 and gave me the opportunity to present some of it to his students at the University of North Carolina School of the Arts (UNCSA). Yingchen Kwok provided feedback on chapter 4, as did attendees at the 2023 NWSA conference. I am grateful to everyone who engaged with my work, as their expertise strengthened the project significantly.

I am beyond thankful for everyone at the University of Washington Press, especially Larin McLaughlin, Banu Subramaniam, and Rebecca Herzig. All three went far above and beyond in shepherding this book through the review and publication process. I am also appreciative to the two external readers for their engagement and feedback. Anonymous review is a deeply undervalued form of labor; I hope the reviewers will see that their efforts led to substantive improvements in the book. I am fortunate that the Wake Forest Humanities Institute provided a book development grant, via funding stemming from a National Endowment for the Humanities (NEH) Challenge Grant, as a result of which I was able to engage the editing services of Brian Conn, who helped me to substantively rewrite the introduction. Any views, findings, conclusions, or recommendations expressed in this book do not necessarily represent those of the NEH.

Finally, I am deeply grateful for the unconditional support of my family, including my parents and brother. My spouse, Daniel Castro, is my rock and biggest supporter. My son, Gabriel Castro-Gupta, and my dog, Pippi, bring the joy into my life that makes it all worthwhile.

ACING SCIENCE

INTRODUCTION

Hegemonic Science, Compulsory Sexuality, and Asexual Possibilities

IN 2009, A STUDY PUBLISHED IN THE JOURNAL *NEUROSCIENCE* USED functional magnetic resonance imaging (fMRI) to compare the brains of "healthy" women to those of women diagnosed with hypoactive sexual desire disorder (HSDD). In their conclusion, the researchers underscored differences between the two groups, describing women without HSDD as "normals," reinforcing the implicit assumption that women who don't want sex are abnormal (Arnow et al. 2009). A decade later, animal behaviorists conducted a series of studies with "noncopulating" (NC) male rats, finding that sexual behavior could be "induced" in them through the administration of certain drugs, as well as the electrical stimulation of specific brain regions. Researchers also argued that "the NC males that have been identified in several species could be equivalent to asexual individuals in humans" (Portillo and Paredes 2019, 5)—implying that the same drugs and electrical stimulation could be used as a medical intervention to induce sexual behavior in asexual humans. In fact, basic scientific research into HSDD has already been translated into clinical applications in humans: in 2015, the US Food and Drug Administration (FDA) approved Addyi, a pharmaceutical treatment for female sexual interest/arousal disorder. "Frustrated that you rarely want sex? It could be Hypoactive (low) Sexual Desire Disorder (HSDD). Fortunately, there's a once-daily treatment known as Addyi, the FDA-Approved 'LITTLE PINK PILL,'" proclaimed the home page of the official website for Addyi in 2023.

In each of these cases, low sexual desire is framed as abnormal, as a condition to be cured rather than simply another way that people may differ. These are just a few examples of the ways in which sexual disinterest is pathologized in hegemonic scientific research, which is, in turn, translated into diagnoses and "treatments" that are applied to people who experience little or no sexual desire. While to some this pathologization may appear innocuous or even helpful, from the perspective of contemporary critical theory, these scientific studies, as well as their translation into medical practice, reflect and reinforce troubling assumptions about sexuality—or asexuality.

What is asexuality? Contemporary asexual communities define an "asexual" or "ace" person as someone who experiences little or no sexual attraction to other people (Catri 2021). Ace individuals and groups contend that sexual disinterest is not necessarily pathological, and that asexual lives can be joyful and fulfilling (Carroll 2024). From this perspective, scientific studies that describe "allosexual" people ("sexual" people, those who experience sexual attraction to others) as "normal" raise red flags. In fact, these and similar scientific studies reflect and can reinforce what asexuality studies scholars call "compulsory sexuality."

The term *compulsory sexuality* was inspired by the term *compulsory heterosexuality*, a concept developed by American lesbian feminist poet and critic Adrienne Rich (1980) to describe the various social and political structures that marginalize lesbian experiences and compel women to participate in the institution of heterosexuality. Drawing inspiration from this term, asexuality studies scholars use the term *compulsory sexuality* to describe the various institutions and norms that marginalize asexual experiences, compelling people to identify as sexual and engage in sexual activity. Compulsory sexuality can have devastating effects on people who identify as asexual, and can restrict possible ways of being-in-the-world for everyone (Emens 2013; Gupta 2015; Przybylo 2019; Mollet and Lackman 2021).

Although there is now a robust body of scholarship on asexuality and compulsory sexuality, this book is the first to undertake a sustained analysis of hegemonic scientific research and its circulation in mainstream US society from these perspectives. Here I am utilizing the term *hegemonic science* as a shorthand to identify a specific knowledge-related set of discourses, practices, and institutions that developed in Europe between 1500 and 1750 and quickly became a global, but still European and white settler–dominated, enterprise, tied closely to intersecting systems of colonialism, anti-Black racism, patriarchy, heterosexism, and ableism.[1] Analyzing hegemonic science and its circulation in mainstream US society from the perspective of asexuality and compulsory sexuality brings to light a number of hidden "pro-sex" biases in this scientific and medical discourse and practice, including the belief that sexual desire is an innate drive; that sexual desire is critically necessary, whether for procreation, pleasure, relationality, and/or national or racial "hygiene"; that all healthy adults experience sexual desire; that the absence of sexual desire in adults is pathological; and that healthy romantic relationships entail sexual intimacy. Informed by these assumptions, hegemonic scientific and medical research in turn reinforces these biases through its findings and their translation into technologies and practices. For example, as will be argued later in the book, in the case of neuroimaging research on sexuality, scientists have often begun with the belief that all people possess a common "sexual response system" that is universal, and that sexual response is distinct from other emotions and motivational states. Studies are designed based on these assumptions, and in turn, their results are used to claim that sexual disinterest is a diagnosable disorder susceptible to medical "treatment," particularly through psychopharmaceuticals (e.g., Arnow et al. 2009).

One of the primary interventions of this book is to expose and challenge the ways in which various fields of hegemonic science reflect and reinforce an essentialized understanding of sexual desire as a universal, innate drive with a socially critical purpose. In turn, this understanding contributes to viewing asexuality as a deficit or a disorder and asexual people as mistaken about or repressing their own sexual desires. Thus, exposing hegemonic scientific assumptions and calling scientific findings about sexual desire into question may undermine their role in reproducing compulsory sexuality in the US. In addition, it may allow for the development of different, nonpathologizing scientific projects about sexual desire, sexuality, and asexuality. Either outcome would contribute to a more ace-inclusive world.

This book goes further, mining hegemonic scientific research on sexuality as a general resource for challenging problematic thinking about sexuality in mainstream US society. Reading this research on sexuality through the analytics of compulsory sexuality and asexual possibilities reveals that categories such as "sexual desire," "sexual pleasure," "sexual activity," and "sexuality" are coherent neither in science nor in the material world that it seeks to explain.[2] For instance, a close reading of scientific findings on animal sexuality suggests that researchers themselves do not have a clear definition of "animal sexual behavior," and that animal behavior cannot necessarily be separated into the categories of "sexual" and "nonsexual." Academics and activists can utilize this insight about the incoherence of the category of "sexual behavior" to further challenge broader social ideas about sex as a distinct activity that is qualitatively different from other types of intimate or pleasurable activities.

Thus, the second main intervention of this book is to demonstrate that reading hegemonic science through the analytics of compulsory sexuality and asexual possibilities productively destabilizes the socially constructed boundary between the "sexual" and the "nonsexual" in the areas of desire, pleasure, and activity. In turn, revealing the incoherencies in these categories creates space to develop new ways of talking and thinking about our desires, pleasures, activities, and relationships that do not consider their "sexual" or "nonsexual" status to be their most important characteristic. Developing these new ways of talking and thinking can play a role in the construction not only of a more ace-friendly world but of one that supports a wider range of possibilities for everyone.

In the remainder of this chapter, I offer an introduction to contemporary asexual identities and the concepts of compulsory sexuality and asexual possibilities, which were developed by feminist and queer scholars working in the nascent field of asexuality studies, and which serve as the primary analytical tools of the book. Then, I introduce its primary object of analysis—hegemonic scientific studies of sexuality—as well as the specific case studies from the science of sexuality that appear in the remaining chapters.

Asexual Identities, Asexual Possibilities, and Compulsory Sexuality

The concepts of compulsory sexuality and asexual possibilities come from the field of feminist and queer asexuality studies, which itself developed out of the lived experiences and activist organizing of people who see asexuality not as a disease but as a sexual identity. The political coalescence of this identity category gained steam in the late 1990s via online networking: people who did not experience sexual attraction or desire or who were not interested in having sex with others began to share their experiences online and come together to form virtual communities, where the work of defining asexuality as a sexual identity took place.[3] One of the earliest, the Asexual Visibility and Education Network (AVEN), founded in 2001, defined an asexual (or ace) person as "someone who does not experience sexual attraction," and this definition has endured, with some modifications, to the present day (Hinderliter 2009). Originally drawing members primarily from the United States, Canada, and the UK, ace internet hubs have expanded geographically, and there now exist a number of region-specific ones for people who identify as asexual.[4] Asexual identity has also moved offline, as ace community members regularly march in Pride Parades and organize meetups.[5]

Contemporary transnational asexual communities tend to define asexuality based on desire—specifically describing asexual people as experiencing little or no sexual attraction to others—and understand asexuality to be a sexual identity and/or orientation (Scherrer 2008; Gupta 2017).[6] These groups also tend to distinguish between sexual attraction and arousal, suggesting that many asexual people experience the latter but not the former.[7] They distinguish too between sexual and romantic attraction, arguing that people may experience one, both, or neither; those who do not experience romantic attraction have adopted the term *aromantic* (or *aro*) as an identity label (Antonsen et al. 2020).[8] AVEN and related sites specifically do not define asexuality on the basis of behavior, asserting that some asexual people may engage in sexual activity while others may not (Carrigan 2011).

Asexual individuals and groups have brought attention to the ways in which they experience marginalization, stigma, and violence as a result of social norms about sexuality. A study of a global sample of 12,449 people on the asexual spectrum (aspec), reported in 2023, found that 64.8 percent of respondents had experienced minority stress, such as verbal aggression, victimization, or health care discrimination, based on sexual or romantic orientation. In addition, approximately 32.2 percent reported suicidal ideation, 10.6 percent had suicide plans, and 2.7 percent had attempted suicide in the previous twelve months (Chan and Leung 2023). A variety of quantitative and qualitative studies have found that aspec people report experiencing stigma, discrimination, invisibility, pathologization, unwanted sex, and "corrective" rape. Aspec people also report experiencing acephobia even in LGBTQ+ spaces (Zheng and Su 2022; Mollet and Black 2023; Brandley and Dehnert 2023; Parmenter et al. 2021; Gupta 2017).

Over the past fifteen years, theorizing in asexual spaces, in addition to the lived experiences of aspec people, has given rise to two bodies of scholarship. One comprises social science research that explores asexual community formation and/or identity development.[9] The second, in which this book is situated, comprises feminist and queer analyses of the political implications of asexuality-as-identity and asexuality-as-concept or lens of analysis. Both bodies of scholarship have been dominated by academics from the US and Europe, although there has been a growth in recent years of work in asexuality studies by those from other parts of the world.[10] Two concepts developed in feminist and queer asexuality studies serve as the primary analytical tools of this book: *asexual possibilities* and *compulsory sexuality*.

Feminist and queer thinkers have developed conceptualizations of asexuality that are inspired by but differ from ace understandings of asexuality as an identity or orientation. For instance, Ela Przybylo and Danielle Cooper (2014, 298) argue for the need to expand beyond an identity model of asexuality. Their "blurrier" understanding of asexuality can include, for example, celibacy and the "political asexuality" of some radical feminists during the US women's liberation movement, and rather than looking for asexuality-as-identity in literary, historical, and other archives, they seek out "asexual resonances" through a "queerly asexual reading strategy." They write, "Such a queer broadening of what can 'count' as asexuality, especially historically speaking, creates space for unorthodox and unpredictable understandings and manifestations of asexuality."

Other feminist and queer asexuality studies scholars have elaborated on this claim, maintaining that asexuality can serve as an analytical or epistemological tool through which different kinds of nonsexual practices, feelings, relationships, and even narrative structures can be identified and not automatically be understood as instances of lack, failure, or dysfunction. This line of argumentation does not claim that all forms of asexuality are always "good" but only that asexuality *can* be good, and must be evaluated contextually (for further development of these ideas, see Hanson 2014; Gupta 2015; Przybylo 2019; Cerankowski 2021; Snaza 2020). Academics have employed a variety of terms for this nonidentitarian understanding of asexuality, including "asexuality as a queer orientation to sexuality" (Snaza 2020), "asexual assemblages" (Kenney 2020), "asexuality as theoretical concept" (Kurowicka 2014), "asexual pleasures" (Cerankowski 2021), and "asexual erotics" (Przybylo 2019). In this book, I use the term *asexual possibilities* to indicate this expansive notion of asexuality as a nonidentitarian concept that can point to various forms of nonsexual activities, feelings, relationships, politics, and so on. Although feminist and queer asexuality studies arose specifically from relatively recent asexual identity movements, concepts such as asexual possibilities can point to other forms of asexuality or nonsexuality, such as virginity pacts, involuntary celibacy (incels), or religious vows of chastity or celibacy, each of which must be evaluated contextually.

The concept of asexual possibilities dovetails with the second major concept elaborated in feminist and queer asexuality studies—namely, the understanding of stigma against people who identify as asexual as part of a broader system of normalization and oppression that simultaneously privileges sexual identities, desires, and relationships and marginalizes asexual and nonsexual ones. Feminist and queer asexuality studies scholars have utilized a variety of terms to name this system, including *sex/sexual normativity*, *the sexual assumption*, and *compulsory sexuality* (Carrigan et al. 2014; Miller 2017; Carrigan 2011; Emens 2013; Gupta 2015). This conceptualization of asexual stigma as part of a broader system of social regulation expands on the concept of "heteronormativity," which was developed by feminist and queer thinkers to describe not only stigma or hatred directed at nonheterosexual people (homophobia) but also the privileging of heterosexuality and the belief that it is the only normal or natural sexual orientation or identity (Morgensen 2021; Warner 1993).

In this book, I use the term *compulsory sexuality*. This specifically draws on the concept of "compulsory heterosexuality," which was developed by American lesbian feminist poet and critic Adrienne Rich in 1980. In brief, Rich (1980) defined the term as the patriarchal and heteronormative structures that impose heterosexuality on women, thereby rendering their sexuality available to men. Compulsory sexuality, then, refigures Rich's notion to emphasize the ways in which social structures impose sexuality on everyone, thereby rendering their sexuality available to others. In an earlier publication, I defined the term as "the assumption that all people are sexual; the norms and practices that compel people to experience themselves as desiring subjects, take up sexual identities, and engage in sexual activity; and the norms and practices that marginalize various forms of nonsexuality (including a lack of interest in sex, a lack of sexual activity, or a disidentification with sexuality)" (Gupta 2015, 134–35). Ela Przybylo (2022, 6) describes compulsory sexuality as the assumptions "that sex is needed for intimacy and healthy romantic relationships, that sex is needed for a person to be healthy and whole, and that one must identify in terms of sexuality in order to be legible to others." Compulsory sexuality underlies discrimination against people who identify as asexual, and it also regulates the behavior of all people, asexual and allosexual alike, rendering asexual and nonsexual feelings, activities, and relationships suspect or inferior (Przybylo 2022; Gupta 2015; Emens 2013; Barounis 2014; Mollet and Lackman 2021).

More recently, some members of ace and aro communities, as well as feminist and queer asexuality studies scholars, have begun utilizing the term *allonormativity* to describe the privileging of those who experience sexual *and* romantic attraction over those who identify as asexual or aromantic (Mollet and Lackman 2021). In this book, I continue using the term *compulsory sexuality* because I am focused specifically on sexual, not romantic, attraction.

As a number of feminist and queer scholars of asexuality have argued (e.g., Owen 2018; Kenney 2020; Gupta 2019a; Vance 2018), it is necessary to understand compulsory sexuality and asexual possibilities as intersecting with other systems of privilege, marginalization, and oppression, such as gender, race, class, nation, and ability/disability. Intersectionality is a theory that was developed by activists and academics of color, particularly Black feminists in the US, to describe the ways in which multiple systems of oppression are coconstitutive. According to intersectionality, systems of oppression are not merely additive but combine to produce qualitatively different experiences of privilege and marginalization for individuals and groups who are differently positioned within intersecting systems of oppression (Cho et al. 2013; Hancock 2016; McCall 2005). As discussed below, an intersectional approach suggests that compulsory sexuality and asexual possibilities cannot be understood in isolation from other systems of oppression; conversely, understanding asexuality helps scholars and activists to understand and resist these other systems.

One consequence of intersectionality is that compulsory sexuality affects different gendered positions differently. For example, white, middle-class men who deviate from compulsory sexuality also deviate from norms related to hegemonic masculinity. White, middle-class women who deviate from compulsory sexuality may experience some social acceptance as a result of norms that associate white femininity with sexual passivity, although such women are also more likely to experience violations of their sexual autonomy, as well as epistemic invalidation of their (a)sexuality (Gupta 2019a; Cuthbert

2022; Przybylo 2014). Historically, some trans people seeking hormone therapy or sex- or gender-affirming surgery in the United States felt it was necessary to "perform" asexuality in order to appease medical gatekeepers and access medical services (Meyerowitz 2002, 158–59). Today, a large percentage of ace and aro people identify as trans or gender-nonconforming, and vice versa.[11] Because many aspects of sexuality are tied to gender and vice versa (e.g., sexual identity is often defined in part based on one's gender and that of the person(s) to whom one is attracted, and hegemonic masculinity is partly defined as sexual assertiveness), disidentification with sexuality may foster disidentification with gender as well (see Gupta 2019a for a discussion).

Regarding disability, asexuality itself has been pathologized as a mental disorder, implying a connection between the two. However, disability and ace communities have not always found common ground in opposition to compulsory sexuality and ableism. People with disabilities are regularly "desexualized" by US society—viewed as childlike and denied access to sexual health education and services, erotic material, and privacy for sexual activity (Kafer 2020; Shakespeare 2000).[12] As a result, some disability studies scholars and activists have argued that people with disabilities are "just as sexual" as people without (see Gupta 2014 for examples). At the same time, some asexual people have distanced themselves from disability on the grounds that asexuality is not a mental or physical illness. This "mutual negation" leaves little space for people with disabilities who identify as asexual. However, in recent years, ace communities have made efforts to be more accessible and to include people who consider their own asexuality to be the result of illness or disability (Kim 2011, 2014; Gupta 2014; Cuthbert 2015; Kurowicka 2023; Lund and Johnson 2014).

The intersection of sexuality with race and colonialism is no less complex. Ideas about sexuality have played a key role in justifying systems of racism and colonialism, and the establishment of those systems in the modern world was intimately tied to European and white settler systems of sexuality. Europeans and white settlers represented themselves as possessing an "appropriate" level of sexual restraint, which was seen as a marker of civilization; in contrast, they represented Black, Indigenous, and other people of color as hypersexual and "savage." In turn, Europeans and white settlers used the supposed sexual "deviance" of people of color as evidence of their inferiority and inhumanity and as justification for their enslavement, colonization, and/

or erasure through genocidal violence (Morgensen 2021; Peiretti-Courtis 2021; Smithers 2017; Schuller 2018; Somerville 2000; Spillers 1987). While many racial and ethnic minority groups were primarily hypersexualized, they have also occasionally been "desexualized" (e.g., the "mammy" is a desexualized stereotype of African American women), and some have been primarily desexualized rather than hypersexualized (e.g., Asian and Asian American men).[13] Thus, notions of "too much" sex *and* "too little" sex have been utilized to position European and white settler sexuality as "just right." Compulsory sexuality is thus about sexuality *and* about race and nation, compelling reproductive sexuality in European and white settler societies while justifying discrimination against external and internal Others who are seen as hypersexual or asexual. As some academics and ace activists have noted, racialized sexual stereotypes make claiming an asexual identity more fraught for those from groups that have been hypersexualized, desexualized, or both (Owen 2018; Rkasnuam 2019).[14]

Europeans and white settlers who displayed too much or "deviant" sexual interest were (and continue to be) pathologized and subject to medical treatment. On the one hand, these people were seen as transgressing the boundaries of whiteness and were grouped with Black, Indigenous, and other people of color on the basis of allegedly shared hypersexuality and sexual deviancy. On the other hand, white sexual deviants were offered a (narrow) path to acceptance through diagnosis and treatment—but this was not available to people of color, creating a hierarchy between white sexual deviants and Black, Indigenous, and other people of color (Patil 2022). As will be argued in chapter 1, this pattern applies to the case of sexual disinterest, as diagnostic categories such as hyposexual desire disorder seem to be primarily reserved for Europeans and white settlers.

In addition to excluding Black, Indigenous, and other people of color from the category of the human by emphasizing their failure to meet white gender and sexual norms, Europeans and white settlers also attempted to colonize Black and Indigenous systems of gender and sexuality. Thus, white gender and sexual categories have been employed as tools of colonial violence, displacing and marginalizing Indigenous systems and concepts related to (a)gender and (a)sexuality (Morgensen 2010, 2012; Lugones 2007, 2010). Indigenous studies scholar Kim TallBear (2022, 22) uses the term *compulsory settler sexuality* to describe the set of norms about gender and sexuality that marks "Indigenous and other marginalized relations as deviant" while attempting to "railroad all of us into rigid relational forms" involving heterosexuality, monogamy, the nuclear family, marriage, and private property. The term *compulsory sexuality* as utilized in feminist and queer asexuality studies thus describes one piece of what TallBear describes as the broader system of compulsory settler sexuality, focused specifically on the systems and norms that privilege sexual attraction or allosexuality over asexuality.[15] As part of compulsory settler sexuality, compulsory sexuality is thus a tool of settler colonial violence. Asexuality, as a European and white settler identity category based on European and white settler understandings of gender and sexuality, also has the potential to be employed as a tool of colonial violence, if it is imposed on Black, Indigenous, and other people of color and if it displaces Indigenous understandings of asexualities and nonsexualities, more broadly conceived. Academics have begun to make connections between broader understandings of asexuality and Indigenous studies concepts such as "sovereign erotics" and "the erotic of abstinence," but more work remains to be done.[16]

Compulsory sexuality also intersects with economic systems of oppression, although these connections also need more academic attention. Compulsory sexuality is one of the ways in which racial capitalism compels reproductive labor, which, in turn, ensures the production and maintenance of the capitalist labor force (Vance 2018; Smith 2020). According to a number of scholars, neoliberal racial capitalism both incites and profits from the endless proliferation of sexual desires, identities, communities, and practices (Hennessy 2000; Curtis 2004; Attwood 2009; Saunders 2020). For instance, corporations market to the "gay consumer," who is imagined as white, cisgender, male, and affluent (Drucker 2015). As another example, "liberated" (white) women are encouraged to pursue empowerment through sexual consumerism, including the purchase of sex toys, erotica, fitness regimens, sexual self-help manuals, clothes, cosmetics, and the like (Attwood 2009). Again, asexuality has the potential to become simply one more niche identity marketed to under neoliberal capitalism, but asexual possibilities also have the potential to trouble capitalism's exploitation of sexuality, if they interrogate the compulsion to reproduce labor or the allure of eroticized advertising, goods, and services.

In this book, I use asexual possibilities and compulsory sexuality as analytical tools through which to read scientific texts, discourses, practices, and institutions. The analytic of asexual possibilities reveals "traces" of asexuality and other forms of nonsexuality as potentially (but not always) fulfilling ways of being-in-the-world, while compulsory sexuality reveals how these objects of analysis privilege allosexuality and marginalize asexuality and nonsexualities. In the next section, I introduce the object of analysis to which I will apply these tools: hegemonic scientific research on sexuality.

The Hegemonic Science of Sexuality

Hegemonic science is merely one of many knowledge-making systems. What then do I mean by *hegemonic science*? There is no single set of theories or methods that distinguishes it from other efforts to investigate the world. Rather, it is a heterogeneous field, best seen as a diverse set of efforts to learn about the world employing observation, experimentation, and theory construction, with a common origin in the scientific revolution that occurred in Europe between 1500 and 1750 and with common institutions (including journals, conferences, universities, laws, regulatory agencies, and funding sources) and discourses (including theoretical paradigms and citational chains) (Okasha 2016; Zack 2014). Hegemonic science has always been a global phenomenon—both in the sense that scientists in the Global North have extracted physical resources, Indigenous knowledge, and objects of study from the Global South and have utilized science as a tool to enforce colonial and racial hierarchies, and in the sense that hegemonic science is also practiced by people in the Global South, sometimes in combination with Indigenous forms of knowledge-production and beliefs about the natural world (Sillitoe 2009; Poskett 2022a, 2022b; Ludwig et al. 2021; S. Harding 1998). In the global enterprise of hegemonic science, Global North countries retain their dominance in terms of scientific output, although Global South countries such as India and China have increased theirs significantly. Yet scientific knowledge produced by researchers in the Global South is routinely underrecognized by hegemonic science (Nakamura et al. 2023; Reidpath and Allotey 2019). In addition, while interest in other (nonhegemonic) knowledge-making systems has grown in the English-language academy, Indigenous and other knowledge-making systems remain marginalized compared to hegemonic science (Held 2023; Tata et al. 2023; Sinclair 2020). Most of the scientific research analyzed in this book was produced by American or European scientists; however, some was produced by researchers in other parts of the world, including China, South Korea, and Mexico.

Although this book focuses on compulsory sexuality in hegemonic science, the latter is not, of course, the only institution—or necessarily the most important—that reflects and reinforces compulsory sexuality. Legal, educational, family, and economic systems, as well as the media and cultural representations, among others, all reflect and reinforce compulsory sexuality. Scholars in feminist and queer asexuality studies have begun to examine many of these institutions through the analytic of compulsory sexuality, although more research remains to be done.[17] This book should be seen as performing one piece of the work of analyzing compulsory sexuality, not as offering a comprehensive examination of every mechanism that enforces it in contemporary US society. Still, science is an important site of institutional control. Although significant segments of the population reject scientific findings on some issues, hegemonic science retains a high status in many societies in the Global North (including the United States) and Global South, in which scientists are considered experts and are consulted regularly on social and political problems. In addition, in many communities, scientific ideas, ways of thinking, and practices permeate daily life, as people are exposed to them via education, the media, and routine interactions with the technologies that result from research (Erickson 2016; Pamuk 2021; Public Face of Science Initiative 2018). Thus, science remains an important site of analysis for those interested in the construction and maintenance of social norms.

This book focuses on hegemonic scientific research on sexuality specifically, which is a prolific body of discourse that emerged in the late nineteenth century in Europe and the United States, and, like hegemonic science in general, quickly became a global enterprise (Fuechtner et al. 2018). Over the past century and a half, such research on sexuality in a number of fields has coalesced under an integrated paradigm. To oversimplify, according to this paradigm, sexual reproduction evolved because it is adaptive (i.e., confers evolutionary advantage) for complex multicellular organisms. Sexual difference, with females producing larger, immobile gametes and males producing smaller, mobile gametes, evolved in many plants and all animals because it is also adaptive. Differences in "parental investment" by male and female parents has led to the evolution of sex- and gender-specific types of sexuality, with males on average less discriminate about sexual activity because they invest less in reproduction, and females on average more discriminate because they invest more. These differences have a genetic basis, as do other aspects of sexuality. In mammals, genes shape sexuality through the action of hormones, first during early development (in utero), when they organize the brain in certain ways, and then, from maturity onward, as circulating hormones influence sexual behavior (Bell 1982; Trivers 1972; Gooren 2006; Jordan-Young 2010).

Many scientists of sexuality now argue that different sexual orientations (hetero-, homo-, and bi-) have a genetic component, leading to differences in prenatal hormone exposure and thus brain structure and function in areas related to sexuality. Researchers have proposed various evolutionary theories to explain homosexual behavior, and have conducted genetic studies of homosexuality. They have also compared prenatal hormone exposure (thought to be reflected by finger-length ratios), physiological response to "erotic stimuli," and brain structure and function between heterosexual and homosexual subjects (Terry 1999; Jordan-Young 2010; Bailey et al. 2016). More recently, some scientists have also begun to see certain types of gender-nonconformity as benign variations, also claiming that "gender incongruence" has a genetic basis, expressed through differences in prenatal hormone exposures that lead to differences in brain structure and function in areas related to gender identity (Kreukels and Guillamon 2016; Erickson-Schroth 2013).

There have been a handful of scientific studies of "asexual orientation" in this same paradigm, mostly by sex researcher Lori Brotto and the University of British Colombia Sexual Health Laboratory, which she directs. These have compared response to erotic stimuli between asexual and allosexual subjects using measures of genital arousal (Brotto and Yule 2011; Skorska et al. 2023), eye-tracking methods (Milani et al. 2023; Bradshaw et al. 2021; N. Brown et al. 2021), and fMRI (Prause and Harenski 2014). One study has examined finger-length ratio in asexual people (Yule et al. 2014).[18]

Feminist and queer science studies scholars have offered cogent critiques of this scientific paradigm (Grasswick 2021; Rolin 2021; Weaver and Fehr 2017). They have persuasively argued that hegemonic scientific research on sexuality often reflects and reinforces long-standing stereotypes about masculine and feminine sexuality—namely, that males are sexually assertive and promiscuous, while females are passive and discerning (e.g., Wesner 2019; Ruti 2015). In addition, this research often reflects and reinforces essentialist thinking about sex, gender, and sexual orientation: first, by assuming that sex and gender differences and sexual orientation are innate, biologically determined, and fixed across life spans and sociohistorical contexts; and second, by assuming that sex (gender) and sexual orientation occur in two mutually exclusive fixed categories (male/female and heterosexual/ homosexual, respectively). In turn, binary, essentialist sex and gender categories inform binary, essentialist sexual orientation categories; for example, gay men are thought to be "like" straight women in terms of their sexuality (e.g., Wesling 2022; Dussauge and Kaiser 2012). Feminist and queer academics and activists have asserted that these assumptions can have significant real-world consequences, especially for women and LGBTQ+ people, for instance, in the development of so-called medical treatments to "cure" homosexuality and "fix" women who express too much or too little sexual desire (e.g., Spurgas 2020; S. Diamond 2017). These thinkers have also argued that the white supremacist, settler imperatives discussed earlier continue to inform scientific understandings of sexuality (Patil 2022; Subramaniam et al. 2016); thus necessitating an intersectional lens of analysis.

Finally, feminist and queer scholars who have analyzed science, including the science of sexuality, have utilized its findings to productively denaturalize social concepts of sex, gender, and sexuality (Tuana 2021; Hird 2009a). As an example of this move, Myra Hird (2006) uses scientific data about sex changes among nonhuman animals to denaturalize human conceptions of trans experiences. She contends that, given the prevalence of sex changes among nonhuman animals, it does not make sense to debate whether trans phenomena are unnatural when they occur in humans.

Building on existing feminist and queer analyses of sexual science, this book uses compulsory sexuality and asexual possibilities as analytics to expose the ways in which hegemonic sexual science reflects and reinforces understandings of sexual desire as an innate drive and contributes to the marginalization of various forms of nonsexuality. After this initial analysis, it rereads scientific findings in order to destabilize normative understandings of sex—especially of sexual desire, pleasure, and activity. Rereading allows for the incoherence of these categories to become clear, which, in turn, can assist in making space for other ways of talking and thinking about the variety of practices, feelings, relationships, institutions, and identities that we currently categorize as "sexual" or "nonsexual."

Before turning to an overview of the remaining chapters of the book, here I offer a brief reflexive statement on my own positionality vis-à-vis this project. Feminist science studies scholars have reasoned that all knowledge is situated, shaped by the specific social, cultural, and historical context in which it is produced and understood (Haraway 1988; Grasswick 2021). This includes the knowledge produced by feminist academics. My position as an academic living and working in the United States, with educational, economic, and racial privilege, has shaped my interest in, experiences with, and attitude toward the elite discourses analyzed in this book. My experiences of trauma resulting from compulsory sexuality have also shaped my approach to this project. The values I bring to it are unabashedly "pro-ace": I have been thinking about and with asexuality since at least 2008, and it is my hope that my academic work on the topic contributes in some way, no matter how small, to a more ace-inclusive world.

Introduction to the Case Studies

This book analyzes four specific areas of hegemonic scientific research on sexuality: the history of the medicalization of hyposexual desire, neuroimaging research on sexual desire, scientific findings on asexual phenomena in nonhuman animals, and scientific research on asexual reproduction. These are slices of the broader field of sexual science, and all generally follow the paradigm outlined in the previous section. They do form distinct clusters, indicated by the citational chains in articles published in each area, but they are also interrelated and mutually influencing.

As these four areas are interrelated parts of the broader field of sexual science, the ordering of the case studies is, to some extent, arbitrary. At the same time, their ordering is intended to demonstrate that even in areas of hegemonic sexual science seemingly distant from understandings of asexuality as identity, compulsory sexuality is both reflected and reinforced. To that end, the case studies are ordered in terms of their "distance" from the identity-based definition of asexuality as characterizing a person who lacks sexual attraction or desire. That is, the case studies move from a lack of sexual desire in humans to "sexual response" in humans to a lack of sexual behavior in nonhuman animals to nonsexual reproduction. The latter case studies build on the understanding of compulsory sexuality developed in earlier ones, applying it to areas increasingly remote from human (a)sexuality.

I start with the history of the medicalization of hyposexual desire, because this is the area of sexual science that interfaces most directly with humans. Chapter 1, titled "'There Is a Great Deal of Denial in This Population': Euro-American Medicine and Compulsory Sexuality," draws on both primary and secondary sources to analyze European and American diagnostic categories that have been developed to describe people who experience little or no sexual attraction. At various points from the late nineteenth century to the present, medical and mental health professionals defined a lack of interest in sex as a disorder requiring treatment. Various terms, from *sexual anesthesia* to *frigidity* to *hypoactive sexual desire disorder*, have been used to describe this so-called illness. Reading this research through the analytics of compulsory sexuality and asexual possibilities reveals that this area of praxis simultaneously relies on and constructs sexual desire as an innate "drive," which, in turn, promotes an understanding of sexual disinterest as a deficit or a disorder. This case study is offered first not only because it is the closest to asexual identity but also because it provides the most direct evidence of the harms of compulsory sexuality in hegemonic science to asexual individuals and communities, thus establishing the stakes of later chapters. The three areas of research examined in the following chapters do not directly involve people as patients to be treated but rather provide the body of "basic" scientific research by which diagnostic categories are justified.

The second case study is examined in chapter 2, "Sex in the Machine: The Neuroimaging of Sexual Desire," which offers an original analysis of neuroimaging studies and neurological models of sexual arousal and desire. I place this case study second, as this area of research remains within the realm of the "human" and focuses on the concept of sexual desire and attraction—which is the key to "asexuality-as-identity" definitions of asexuality.[19] This case study was chosen in part because, in recent years, neuroimaging research has captured the public's attention and has helped to shape debates about what it means to be human, thus making it an important area to examine from a justice-oriented perspective. At the same time, a critical reading of this research enables a disruption of the categories of sexual desire, arousal, and pleasure. Reading neuroimaging studies on arousal and desire through the analytics of compulsory sexuality and asexual possibilities reveals the ways in which this research relies on and reinforces essentialized understandings of sexual desire by focusing on it as an innate drive, and by insisting on drawing boundaries between it and other emotional and motivational states. At the same time, however, rereading this research also demonstrates that sexual desire, arousal, and pleasure are not distinct categories, either in science or the material world it seeks to capture, thus clearing the way for more flexible and socially situated understandings of desire and attraction.

The third case study is examined in chapter 3, "Pandas, Voles, and Rams: Asexual Phenomena in Nonhuman Animals," which examines scientific findings on sexual (non)activity among nonhuman animals through the analytics of compulsory sexuality and asexual possibilities. I place this case study third because it moves from the category of the human to that of nonhuman animals. In addition, it focuses on sexual (non)behavior, which is not part of the definition of asexuality-as-identity but can fall under the broader umbrella of asexual possibilities. This case was chosen in part because scientific research on the sexuality of nonhuman animals has played a role in public debates about sexuality, with some scientists and activists claiming that evidence for "homosexuality" in nonhuman animals proves that homosexuality in humans is "natural." In addition, this case study allows for a disruption of the category of "sexual activity." The chapter first utilizes Angela Willey's work on compulsory monogamy in research on voles to demonstrate the influence of compulsory sexuality in this area of research. It then offers an original analysis of scientific findings on sexual (non)behavior in other nonhuman animals, including guinea pigs, rams, and rhesus monkeys. The chapter argues that the scientific research on sexual (non)behavior in nonhuman animals reflects and reinforces compulsory sexuality through its limited scope, focus on pathology, and use of pejorative language (e.g., the use of the term *duds*); however, evidence of asexual phenomena in nonhuman animals can also serve to normalize asexual phenomena among humans. At the same time, as mentioned above, rereading this research through the analytics of compulsory sexuality and asexual possibilities demonstrates that, like sexual desire, arousal, and pleasure, the concept of sexual activity is not a self-evident category, either in science or the material world it seeks to describe.

The fourth case study is examined in chapter 4, "Amoebas Are Us? Asexual Reproduction and the Category of Sex," which examines hegemonic scientific research on asexual reproduction through the analytics of compulsory sexuality and asexual possibilities. This case study moves even further beyond the category of the human to that of all living organisms. In addition, it focuses on asexual reproduction, which is even more distant from the definition of asexuality-as-identity than sexual (non)behavior but again can fall under the broader umbrella of asexual possibilities. This case study was chosen, in part, because human asexual communities have both identified with and distanced themselves from nonhuman organisms that reproduce asexually—either finding commonality with amoebas or drawing a boundary between human asexuality, which they define as low sexual attraction and desire, and nonhuman asexuality, which is instead associated with asexual reproduction. At the same time, this case allows for a disruption of the category of "sexual reproduction." The chapter offers an overview of scientific discussions of asexual reproduction, along with different scientific explanations for the evolution of sex and sexual reproduction. Reading this research through the analytics of compulsory sexuality and asexual possibilities reveals a strong "pro–sexual reproduction bias" in this scientific literature but also identifies instances in which the scientific discourse about (a)sexual reproduction calls compulsory sexuality into question. This chapter also utilizes compulsory sexuality and asexual possibilities as analytics to examine feminist and queer academic engagements with bacterial (a)sex and asexual reproduction. In general, feminist and queer studies scholars have claimed that bacterial (a)sex and asexual reproduction provide further evidence for the diversity of sex and reproduction in nature, or, in other words, for the "queerness" of nature. This chapter maintains that although this approach has much to recommend it, it can also reflect and reinforce compulsory sexuality. I argue instead that this research can be employed not as evidence of a "queer" nature but as a way of demonstrating the incoherency of the category of sexual reproduction and the overarching category of "sex."

The conclusion of the book draws from the previous chapters a general critique of the way sexual desire is understood in much of science as an essential drive, an understanding that leads to harm against asexual and nonsexual people. The conclusion also draws from chapters 2 through 4 the argument that rereading scientific research on sexuality can desentiment and reveal incoherencies in the categories of sexual pleasure, activity, desire, and reproduction, as well as the overarching category of sex itself. These categories are no more coherent entities-in-the-world than are gender, race, disability, or any other social construct. Unsettling these concepts makes space for (1) seeing the absence of sexual desire neither as deficit nor dysfunction; and (2) developing new ways of talking and thinking about all of the desires, relationships, practices, and identities that US society currently categorizes as sexual or nonsexual. The conclusion asserts that both moves assist in the creation of a more ace-friendly world, which, simultaneously, is one that supports a greater diversity of human and nonhuman lives.

CHAPTER I

"There Is a Great Deal of Denial in This Population"

Euro-American Medicine and Compulsory Sexuality

AT SOME POINT IN THE 1960S OR 1970S, A MARRIED COUPLE, "DONALD" and "Donna," sought therapy from Helen Singer Kaplan for Donald's erectile difficulties and Donna's lack of desire for sex. According to Kaplan, Donald's issue was the result of Donna's, but the cause of hers was initially unknown. Over the course of the sessions, Kaplan concluded that Donna's lack of desire was due to her hostility toward Donald. As stated in the case report, through therapy, Donna "was made to face the fact that no self-respecting man would put up with her outrageous behavior and that she was in serious danger of losing her marriage" (Kaplan 1979, 112).

Donna initially resisted this analysis, but eventually came to agree with Kaplan. As the latter explained, "the patient saw that she was 'doing it to herself' and that this was to her disadvantage" (112). In the therapist's view, the case of Donald and Donna provides evidence that although "inhibited sexual desire" is difficult to treat, it can be successfully addressed in some cases through lengthy psychosexual therapy. Overall, Kaplan concluded, "desire problems constitute a separate clinical subgroup" and "patients suffering from blocked desire have, as a group, deeper and more intense sexual anxieties and/or greater hostility towards their partners and/or more tenacious defenses than those patients whose sexual dysfunctions are associated with erection and orgasm difficulties" (xvii).

Kaplan's work with couples seeking treatment for "inhibited sexual desire" is but one example from a long history of European and American medical and mental health professionals approaching sexual disinterest as a disorder in need of treatment. This chapter draws on both primary and secondary sources to analyze European and American diagnostic categories that have been developed to describe people who experience little or no sexual attraction. Since the late nineteenth century, various terms, from *sexual anesthesia* to *frigidity* to *hypoactive sexual desire disorder*, have been used to describe this so-called illness.

Reading this research through the analytics of compulsory sexuality and asexual possibilities reveals that this area of scientific and medical praxis has conceptualized sexual desire as serving some necessary and valuable purpose, whether that was to encourage reproduction, produce pleasure, sustain pair-bonding and families, and/or maintain national or racial "hygiene." As a result, it has promoted the idea that among healthy adults, sexual desire should be universally present; if absent, it is a sign of disorder necessitating treatment. While these diagnostic categories may have been helpful for some people, they have also promoted a stigma against those who are sexually disinterested, impeding the acceptance of this as a benign sexual variation. Below I introduce a few key issues in the study of this history before turning to an analysis of the diagnostic categories themselves.

Approaching a History of European and American Diagnoses for Sexual Disinterest

This chapter utilizes compulsory sexuality and asexual possibilities as analytics through which to examine the various European and American diagnostic categories that have been developed since the late nineteenth century to apply to people who experience little or no sexual attraction.[1] The pathologization of sexual disinterest has occurred in the context of a broader pathologization of nonnormative forms of gender and sexuality, including homosexuality (e.g., Terry 1999), nymphomania (Groneman 2001), sex addiction (Irvine 1993), transsexuality and transgender experiences (Shuster 2021; Gill-Peterson 2018; Malatino 2020; Meyerowitz 2002), masturbation (Laqueur 2004), sadism and masochism (Lin 2017), and hermaphroditism and intersexuality (D. Rubin 2017; Reis 2009).

Key themes have emerged from critical scholarship examining this history. First, and perhaps most obviously, the pathologization of nonnormative genders and sexualities has played an important role in their social regulation, encouraging conformance to norms and punishing those who deviate from them, or are seen to do so. In addition, these discourses have played a major role in shaping how people think about and experience their own sexual behaviors, desires, and identities (e.g., Groneman 2001; Terry 1999).

Second, pathologization has not (only) been a "one-way" process through which experts have imposed their own ideas about sexual deviance on society or individuals. Instead, experts are often themselves influenced by broader social norms about gender and sexuality, and, in many cases, have been in dialogue (albeit usually from a position of power) with patients seeking "treatment" for related issues (e.g., Meyerowitz 2002; Oosterhuis 2000). In some cases, medical and mental health discourses have been mobilized by individuals and/or communities as harm-reduction strategies (e.g., the strategic use of diagnoses as a way to legitimize gender-affirming medical treatments) and/or as the basis for more radical political projects (e.g., transforming a medical diagnosis into a politicized identity) (A. Johnson 2019; Conrad and Schneider 1992).[2]

Another key point in the academic literature is that the pathologization of nonnormative genders and sexualities has played a role not only in their regulation but also in other, intersecting projects of oppression, particularly those related to disability, class, race, and nation. For instance, sexology, with its heyday at the end of the nineteenth and the beginning of the twentieth century, was closely tied to eugenics, colonialism, and institutional racism.[3] Sexologists constructed Europeans and white settlers, as a group, as appropriately heterosexual, while diagnosing individual members with sexual "pathologies." In contrast, they constructed Black, Indigenous, and other people of color, as a group, as sexually deviant, which in turn was employed as evidence of their inferiority and justification for their enslavement, colonization, and/or erasure through genocidal violence. Early sexologists combined "evidence" from colonial anthropological studies on the hypersexuality of Black, Indigenous, and other people of color with case studies of individual white patients to construct their sexual taxonomies and theories. On the one hand, white sexual deviants (along with white working-class people and white people with disabilities) were associated with

Black, Indigenous, and other people of color, and each of these groups was targeted for eugenic sterilization. On the other hand, white sexual deviants had a limited avenue for "rehabilitation" through diagnosis and treatment, an option not available to nonwhite people. Thus, diagnostic categories developed in sexology simultaneously pathologized white sexual minorities and associated them with the assumed sexual deviance of people of color, while also maintaining white supremacy by suggesting that individual white sexual minorities were salvable in a way that people of color, as a group, were not (Patil 2022; Morgensen 2021; Peiretti-Courtis 2021; Smithers 2017; Schuller 2018; Klesse 2015; Brennan 2015; Somerville 2000). While the connections between the pathologization of sexual deviance and systems of ableism, racism, and colonialism were perhaps more explicit during the late nineteenth and early twentieth centuries, these continue today; for example, contemporary sexualities research on Black women focuses primarily on STIs, HIV, and sexual risk, with little focused on sex counseling, education, or therapy (Hargons et al. 2021). At the same time, American and European sexological categories like "gender identity disorder" have been imposed on sexual and gender-nonconforming people in the Global South, sometimes replacing indigenous understandings of gender and sexuality (Suess et al. 2014).

This chapter adds to the existing genealogical scholarship on the pathologization of nonnormative sexualities by focusing specifically on the history of the pathologization of "sexual disinterest."[4] I draw partly on existing research, particularly Alison P. Moore and Peter Cryle's work on "frigidity" for the early part of this genealogy and Janice Irvine's work on "inhibited sexual desire" for the later part.[5] I also offer original analyses of primary source documents, including publications by Richard von Kraft-Ebbing, Magnus Hirschfeld, and Helen Singer Kaplan. Adding this history to existing ones of sexual pathologization emphasizes that European and American medical professionals in the fields of sexology and sexual health cannot be described as unremittingly antisex or sex-negative. Rather, they were often prosex or sex-positive about certain forms of sexual activity: generally, in the early part of the twentieth century, they supported only white, monogamous, heterosexual, reproductive sexuality, but, at least since the 1970s, they have supported a greater variety of sexual activities, desires, and identities.

As will be demonstrated in this chapter, as in other cases of sexual pathologization, that of sexual disinterest has not been a one-way process; rather, medical and mental health professionals interacted with patients seeking treatment for sexual disinterest, and, in turn, both were encouraged to see it as a problem by broader social trends. Again echoing other cases, the pathologization of sexual disinterest has also been influenced by ideas about gender, class, race, nation, disability, and other social categories. For instance, in some periods, researchers and medical professionals thought men were more likely to suffer from desire disorders, while in others, women were more likely to be so diagnosed (Cryle and Moore 2012; Irvine 1993). In regard to race and coloniality, as will be argued, while "sexual restraint" was viewed as a positive characteristic for Europeans and white settlers (Owen 2018), "too much" among white people was seen as a problem, due to fears of racial suicide. Thus, the pathologization of sexual disinterest in the late nineteenth and early twentieth centuries was part of a biopolitical project to ensure the futurity of white nation-states. At the same time, while sexologists were unlikely to diagnose individual people of color with hyposexuality, some minoritized groups (such as Asian and Asian American men, people with disabilities, and older people) were pathologized as "hyposexual."[6] Although more research is needed, it is likely that sexually disinterested white people were simultaneously linked to desexualized groups and elevated above them through the possibility of diagnosis and cure.[7]

Before moving to the genealogy itself, a few clarifications: First, the history of European and American diagnostic categories for sexual disinterest and the history of "asexual identities" are overlapping but not identical. Since the late nineteenth century, diagnoses have been applied to people who have reported a lack of sexual desire or interest in sexual activity to medical professionals. For some, this absence may have been an (important) aspect of their lives or their identities, but prior to the development of asexual communities in the late 1990s and early 2000s, it is unlikely that many would have experienced their (a)sexuality or identified as asexual in the way that some do in the contemporary period.[8] It is also likely that many who experienced sexual disinterest were never diagnosed with a sexual desire disorder.

Second, I am not suggesting that "sexual disinterest" is a transhistorical category. Rather, within a specific intellectual tradition (hegemonic sexual science and medicine since the late nineteenth century), a diagnostic category related to a lack of sexual desire or libido has cropped up repeatedly, but understandings of it have changed over time. For example, at some points, it was conceptualized primarily as a psychological problem, while at other points it was understood to be an organic or physiological disorder. Over time, as well, the universe of people who could be diagnosed with a desire disorder has expanded. In the early part of the twentieth century, homosexuality itself was considered pathological in American and European medical practice, but by the 1970s, in some therapeutic circles, lesbians and gay men could be diagnosed with a desire disorder if they were not sexually interested in same-sex partners. As the sexual desires of groups previously deemed "asexual" by society (e.g., older people and those with disabilities) began to be recognized, they became part of the universe of people who could be diagnosed with a desire disorder. I hope to demonstrate in this chapter, however, that there are also remarkable similarities between, for instance, Magnus Hirschfeld's early twentieth-century understanding of "aneroticism" and Helen Singer Kaplan's late twentieth-century understanding of "hypoactive sexual desire."

Finally, I am not conducting a genealogy of the term *asexuality*, which has been utilized in different ways in American and European medical discourse over the past 150 years, most commonly in the context of "asexual reproduction" (which is discussed in chapter 4). Occasionally, *asexual(ity)* has been used by scientists or doctors to refer to people who do not experience sexual attraction or desire, sometimes in pathologizing ways, sometimes more neutrally, and I note cases of this below. With these clarifications in mind, I now turn to a genealogy of the pathologization of sexual disinterest.

European Ideas About Sexual Disinterest Prior to the Late Nineteenth Century

Certainly sexual disinterest was a concern in European and later white-settler societies long before the nineteenth century, a fact attested to by the existence from antiquity to the present of literature recommending different aphrodisiacs (McLaren 2007).[9] In early modern Europe, men were sometimes accused of frigidity or impotence in annulment trials in ecclesiastical courts. These terms could refer to a lack of interest, inability to experience pleasure, or inability to engage in intercourse and/or an inability to produce semen (or "seed"). Frigidity was thought to be the result, in some cases, of supernatural causes (e.g., witchcraft) or, more commonly, natural causes (e.g., constitutional coldness) (McLaren 2007; Cryle and Moore 2012). Interestingly, infertility alone was not generally grounds for annulment if the sex act could be completed, indicating that the ability to engage in sex was possibly considered more fundamental to marriage than reproduction. With notable exceptions, most European ecclesiastical thinkers in this time period were not concerned about frigidity or impotence in women (except when a physical impediment prevented copulation, which was rare), because they saw women as capable of participating in copulation without desire or pleasure or the production of seed (Cryle and Moore 2012). In addition, these thinkers often assumed women were more licentious than men, so they did not think that women were likely to "suffer" from this problem (Philip 2019). Marion Philip (2019) offers a reading of Parisian ecclesiastical frigidity cases between 1665 and 1789 through the lens of contemporary asexual identities, focusing on the case of Jean Joseph Soucany, who (unusually) admitted to the court that he was not sexually interested in his wife. Philip points out the difficulty of interpreting the case, as Soucany's statement of disinterest could be the result of a variety of factors, but affirms that it marked him as nonnormative for his time. In general, the ideas about frigidity and impotence developed in the ecclesiastical courts were influential in shaping later European medical approaches to sexual disinterest (Cryle and Moore 2012).

Beginning in the 1780s, there was a gradual shift in European medical discourses about frigidity, as doctors and medical writers began to recognize its existence in women as well as men. *Frigidity* could still be used to mean disinterest in sex, inability to experience pleasure during sex, and/or inability to engage in sex. By the late nineteenth century, there was a proliferation of publications on frigidity and other related diagnoses by doctors, sexologists, and, eventually, psychoanalysts in Europe and the United Sates (Cryle and Moore 2012).

The Late Nineteenth to Early Twentieth Centuries: European Sexologists and Sexual Anesthesia

In the nineteenth century, European and American scientists, medical practitioners, and colonial officials began to apply so-called modern scientific methods to the study of sexual activities and desires, leading to the development of the field of sexology. As noted in the introduction to this chapter, while ostensibly a project about sexuality (and gender), sexology was also closely connected to eugenics, racial science, and colonialism. A number of sexologists included a category specifically for sexual disinterest in their diagnostic systems. Here I focus on the work of two: Austro-German psychiatrist Richard von Krafft-Ebing (1840–1902), one of the most influential early sexologists; and German sexologist Magnus Hirschfeld (1868–1935), an influential "second generation" sexologist and early advocate for sexual and gender minorities (Kościańska 2020).

Krafft-Ebing's major opus, *Psychopathia Sexualis*, was first published in 1886 and contained fifty-one case histories. The twelfth edition, which was the last one he worked on before he died, contained information from over three hundred cases (Oosterhuis 2000, 47). In his writing, Krafft-Ebing divided the sexual perversions into several broad categories: paradoxia (abnormal periods of sexual activity, e.g., in childhood or in old age), anesthesia and hyperesthesia (pathological absence or increase of the sexual drive), and paresthesia (perverse expression of the sexual drive) (Davidson 2001). Krafft-Ebing himself was most interested in the paresthesias, focusing on sadism, masochism, fetishism, and contrary sexual instinct (or homosexuality), but also described many other "conditions" (Oosterhuis 2000, 47).

Krafft-Ebing's category of "sexual anesthesia" served as a diagnostic category for sexual disinterest or low sexual motivation. Two ideas underlay his decision to view sexual disinterest as pathological: First, he understood the "sexual instinct" to be a biological phenomenon, theorizing that it arose from the nervous system and a particular center in the cerebral cortex, possibly next to the center for olfaction. Therefore, "disturbances" in the sexual instinct could be understood as a biological dysfunction. At the same time, he understood it to be a psychological phenomenon with a "natural" purpose. Early on in his career, he believed that its natural purpose was reproduction, and therefore that phenomena that deviated from this (e.g., homosexuality and sexual anesthesia) were disturbances or diseases of the sexual instinct (Davidson 2001; Oosterhuis 2000). As suggested by Jacinthe Flore (2020), while most academics have examined Krafft-Ebing's treatment of homosexuality, "the drive of which Krafft-Ebing spoke at length was not solely a question of object choice. . . . An understanding of drive as subject to amount, degree and balance circulates in the work of Krafft-Ebing and his contemporaries" (31).

In addition to viewing sexual disinterest as a disorder, Krafft-Ebing also viewed sexual anesthesia as evidence of an antisocial personality. Several ideas led him to this interpretation. For one, in addition to reproduction, he also identified a larger constructive role for sexuality (Oosterhuis 2000), writing: "Sexual life is no doubt the one mighty factor in the individual and social relations of man that discloses his powers of activity, of acquiring property, of establishing a home, and of awakening altruistic sentiments toward a person of the opposite sex, toward his own children, as well as toward the whole human race. Sexual feeling is really the root of all ethics, and probably also of aestheticism and religion" (Krafft-Ebing 1906, 1–2). He especially saw sexuality as an important, indeed essential, part of a loving relationship. In fact, as he came to place more value on affection (and not simply reproduction) as the "natural" purpose of sexuality, his attitudes toward homosexuality softened.[10] While this evolving view may have led Krafft-Ebing to adopt a more tolerant attitude toward some types of sexual expression in his later years, it simultaneously supported his belief in the pathological nature of sexual anesthesia (Oosterhuis 2000).

Krafft-Ebing also saw sexuality as the best expression of an individual's personality. According to Arnold Davidson (2001, 21), nineteenth-century European psychiatrists, including Krafft-Ebing, saw sexuality as revealing the "individual shape of the personality." Krafft-Ebing (1905, 81) wrote, "The [sexual] anomalies are very important elementary disturbances, since upon the nature of sexual sensibility the mental individuality in greater part depends." These two ideas—that sexuality was necessary for affectionate relationships and that it reflected an individual's entire personality—led Krafft-Ebing to view sexual anesthesia as evidence of a more generalized antisocial personality (Davidson 2001).

Krafft-Ebing discussed a number of case histories of sexual anesthesia in his publications; a 1906 English translation of *Psychopathia Sexualis* includes eight. He classifies the first six of these (five men and one woman) as "congenital" sexual anesthetics, whom he describes as "functionally sexless individuals" with "degenerative defects" and "other functional cerebral disturbances, states of psychical degeneration, and even anatomical signs of degeneration" (1906, 61). This discussion of "degeneration" reveals the eugenicist, racist, and colonialist politics of sexology. In general, many early eugenicists saw "degenerative defects" as hereditary and believed that allowing white "defectives" to reproduce would lead to a degeneration of the white race. They advocated for programs, such as forced sterilization, to prevent them from reproducing (Brennan 2015; Roll-Hansen 2010; Kościańska 2020). Thus, a diagnosis of sexual anesthesia may have placed a person at risk of eugenic targeting.

The first three of the cases of congenital sexual anesthesia (all men) appear to have been Krafft-Ebing's own patients. The first (K.) consulted Krafft-Ebing for advice about his "abnormal sexual condition." It is not clear why the second (W.) saw Krafft-Ebing, while the third (P.) sought help as a result of "spastic spinal paralysis." Yet, although all three appear to have sought help, they do not seem particularly troubled by their absence of sexual interest. Of K., Krafft-Ebing (1906, 62) wrote, "Excepting this want of sexual instinct, K. considered himself quite normal. No psychical defects could be detected. He was fond of solitude, but of a frigid nature, without interest in the arts or the beautiful, but a highly efficient and esteemed official." Of W., he wrote, "His deficiency did not seem to cause him any worry," and of P., "Patient did not feel this absence of sexual sensation" (62–63). Krafft-Ebing borrows the remaining three cases of congenital sexual anesthesia (two men and one woman) from William Alexander Hammond's 1887 book, *Sexual Impotence in the Male and Female*. In his discussion of contrary sexual instinct (homosexuality), Harry Oosterhuis (2000, 195) points out that Krafft-Ebing often published the letters and words of his patients unedited, even when they criticized the medical model, thus allowing "contemporary readers [to] find subjective experience, dialogue, multivocality, divergent meanings, and contradictory sets of values in *Psychopathia Sexualis.*" This seems to be the case with sexual anesthesia as well; while Krafft-Ebing presents his sexual anesthetics as pathological, he also allows them to represent themselves as nonpathological variations from normality.

After presenting his six cases of true congenital anesthesia, Krafft-Ebing (1906, 64) discusses two, much longer "transitional cases," who, he says, may possess a weak "mental side of the *vita sexualis*" but may have undermined this through masturbation before it could develop fully. According to him, when the early sexual element is undermined, an "ethical defect" is manifested. The first case, F. J., was a nineteen-year-old student, who Krafft-Ebing says had no sexual feeling toward the opposite sex and experienced none in the sexual act.[11] F. J. was institutionalized twice, once after attempting suicide and again after a maniacal outbreak. Krafft-Ebing describes F. J. as "destitute of moral and social feelings," lacking emotionality, displaying logical peculiarities, and frivolous, vulgar, cynical, and ironic (65–66). The second case, E., was a thirty-year-old journeyman painter who occasionally experienced sexual desire but "never for a natural gratification of it." E. did not want any more children to be brought into the world—he despised women for their reproductive capacity, considered self-castration, and was arrested while attempting to mutilate a boy's genitals. Krafft-Ebing describes E. as mentally abnormal, violent, irritable, selfish, and weak-minded, adding that "social feelings were absolutely foreign to him" (66–67). These two "transitional" cases are very different from the congenital ones: the patients exhibit more pathological behaviors (including violence toward themselves and others), and Krafft-Ebing is much harsher in his judgment of them. It is possible those with "congenital anesthesia" were Krafft-Ebing's private patients, who tended to be middle- or upper-class, while he may have encountered the two with "transitional anesthesia" in psychiatric or penal institutions (Oosterhuis 2000).

Before discussing acquired anesthesia, Krafft-Ebing notes that the cases he has presented so far demonstrate that anesthesia may sometimes be tied to genital deformities or diseases, but not in every case. He also characterizes frigidity as a milder form of anesthesia and as more common in women than in men. He describes the signs of frigidity as "slight inclination to sexual intercourse, or pronounced disinclination to coitus without sexual equivalent, and failure of corresponding psychical, pleasurable excitation during coitus, which is indulged in simply from sense of duty," adding that he has often heard complaints from husbands about their wives' frigidity (1906, 68).

In his last section on sexual anesthesia, Krafft-Ebing briefly describes "acquired anesthesia," but he offers no case histories. According to him, an "acquired diminution" of the sexual instinct or libido may be the result of psychological or physiological factors, or some combination. Causes include (among others): aging, education, intense mental activity, depression, sexual continence, sexual excess, castration, disturbances in nutrition, atrophy of the genitals or genital glands, diseases in the spinal cord or brain, hysteria, and insanity (1906, 68–69). As we will see later in the chapter, in some ways Krafft-Ebing's definition of acquired anesthesia is very similar to twenty-first-century definitions of hypoactive sexual desire disorder: he suggests that hypoactive desire is a variable phenomenon, and that people may experience a decrease in sexual motivation at different points in their life for a number of complicated reasons, including biological, psychological, and social factors.

Magnus Hirschfeld, an influential "second generation" sexologist and early advocate for sexual and gender minorities, discusses "sexual anesthesia" most extensively in his three-volume work *Sexualpathologie* (Sexual pathology), published in 1916, 1918, and 1921 (later translated into English).[12] The third volume includes a section on "impotence," which Hirschfeld divides into several different subcategories. Among these he includes "lack of the sexual urge (libido deficiens)," which he says is a widespread condition in both men and women. He classifies "lack of the sexual urge" as one type of "cerebral impotence" and also uses the terms *anerotismus* (aneroticism) and *asexualität* (asexuality) to describe this disorder (1932, 250; 1921, 162). When describing aneroticism, Hirschfeld begins by arguing that many who claim to "have no sexual needs" are often hiding socially inappropriate sexual desires and/or have found ways to repress their urges. He goes on to write that a lack of the sexual urge may be caused by "physical factors," including stressful situations, mental exhaustion, alcohol, drugs, and/or various forms of physical and mental illness, as all of these can affect "libido-causing substances" (hormones). Hirschfeld claims that physically caused aneroticism is usually temporary and can be cured, although a complete lack of libido-causing substances can be difficult to treat; still, even in those cases, implantation of "sexual gland tissue" is sometimes successful.[13] He concludes this section by discussing what he calls "psychically conditioned defects of the urge." Unlike physically conditioned defects, psychically conditioned ones, he says, are less serious for women because they fulfill their marital duties "more or less passively." He also asserts that psychically conditioned defects are never absolute, only relative—meaning a person is only anerotic toward specific types of people or in specific situations; for example, some are only attracted to their own person (automonosexualism) or only to the same gender or sex (homosexuality). Some people find their spouse to be unattractive or repellant due to specific physical characteristics (Hirschfeld offers the example of a man who was repelled by his wife's false teeth). Others are anerotic because of inexperience or nervousness about sexual performance (here he offers a detailed case study of "Mr. Z." as an example). Hirschfeld (1932, 250–71) claims that many cases of psychically conditioned defects of the urge can be successfully treated through "hypnotic and other psychotherapeutic cures."

In regard to pathological sexual disinterest, the primary difference between Hirschfeld and Krafft-Ebing is that the former does not distinguish between "congenital" and "acquired" sexual disinterest, suggesting that all aneroticism is acquired. While this avoids the eugenic language of "degeneracy," Hirschfeld's discussion is in some ways more pathologizing than Krafft-Ebing's, because in Hirschfeld's view, (almost) all aneroticism is temporary or situational and can be cured by physical or mental health treatments. His work is also interesting because it marks an early instance of the use of the modifier "asexual" in connection with the category of "transvestitism."[14] Most commonly, he argues that transvestites can be heterosexual, homosexual, bisexual, or "automonosexual/monosexual," the last of which is not a synonym for aneroticism (or for modern-day understandings of asexuality) but rather indicates a person who is erotically attracted to their own body or person (1910, 1918, 1991).[15] However, in the second volume of *Sexualpathologie* (1918), of some automonosexual transvestites, Hirschfeld writes, "I knew those who were content if they could go for a walk as a woman from time to time. They had neither erections nor ejaculations nor the need for any kind of sexual intercourse, neither with female nor male persons. They could therefore probably be described as asexual" (145). He also writes, "About 35 percent of transvestites are heterosexual, just as many are homosexual, and about 15 percent are bisexual, while most of the remaining 15 percent are automonosexual, and some are perhaps also asexual" (144–45).[16] As suggested by scholars, Hirschfeld's primary goals in terms of homosexuality and transvestitism were emancipatory (Mancini 2010; D. Hill 2005), so the placing of "asexual" alongside heterosexual, homosexual, bisexual, and automonosexual as a descriptor of the (a)sexuality of "transvestites" suggests a perhaps less pathological understanding of sexual disinterest than appears in his discussion of cerebral impotence.[17]

In addition to Hirschfeld, many other prominent sexologists, including Iwan Bloch (German, 1872–1922), Albert Moll (German, 1862–1939), and Havelock Ellis (English, 1859–1939), accepted, to a greater or lesser extent, Krafft-Ebing's definition of sexual disinterest as a sexual disorder. However, there also seems to be significant diversity in the terms employed to describe it. Ellis, for instance, lists *frigidity*, *hyphedonia*, *anhedonia*, *anesthesia*, *anaphrodism*, and *erotic blindness*, suggesting also that sexologists differed in their understanding of this disorder: Many seem to have included both sexual disinterest and the inability to experience pleasure during sex within the same category, while others separated the two. Sexologists also seem to have disagreed about whether more men or more women experienced sexual anesthesia, what its prevalence was among women, whether there was such a condition as "true" or complete sexual anesthesia, and whether it was more prevalent among the middle and upper classes than the lower classes (Ellis 1936).[18] As noted above, like Krafft-Ebing, many sexologists associated sexual disorders, including sexual anesthesia, with hereditary defects and saw those with them as a threat to "racial hygiene" (Kościańska 2020; Brennan 2015; Klesse 2015). Knowledge generated by colonial officials about the populations they ruled was also incorporated into sexological categories related to sexual disinterest. For example, according to anthropologist Seth Palmer (2014), French colonial officials in Madagascar divided the "sarimbavy" (a category of gender-variant, male-bodied persons) into two groups: sexuals and asexuals. Palmer suggests that Euro-American sexologists, including Havelock Ellis, had read their writing and incorporated it into their own diagnostic systems.[19] Thus, while sexologists were unlikely to diagnose individual people of color with sexual anesthesia, diagnostic categories such as that one were part of the broader institution of scientific racism, which served to legitimate colonial rule and white supremacy (Somerville 2000; Brennan 2015). In the following decades, terms like *sexual anesthesia* and *sexual anhedonia* (diagnoses that could be applied to men or women) became less popular.[20]

The Late Nineteenth to Early Twentieth Centuries: Freud and the Sexual Drive

Although Sigmund Freud (1856–1939) himself was not particularly interested in specific diagnostic categories for sexual disinterest, his work was influential in promoting the idea that sexual desire plays a key role in human psychology and sociality and that sexual repression is the cause of many individual and social ills. In many ways, Freud drew on the research of the sexologists in his conceptualization of drives, and specifically the sexual drive (Sulloway 1992; Oosterhuis 2000). Synthesizing Freud's vast corpus, Laplanche and Pontalis (1973, 214) describe his understanding of a drive (*Trieb*, often translated as instinct): "[a] dynamic process consisting in a pressure (charge of energy, motricity factor) which directs the organism towards an aim. According to Freud, an instinct has its source in a bodily stimulus; its aim is to eliminate the state of tension obtaining at the instinctual source; and it is in the object, or thanks to it, that the instinct may achieve its aim."

For Freud, the sexual drive is one of our primary drives.[21] Laplanche and Pontalis (1973) describe the term *libido* in Freud's work as the psychic representation of the sexual drive. Freud's primary innovation in his conceptualization of the sexual drive and libido was his assertion that it is initially "undirected" (not directed to a specific type of person or activity) but becomes so as a result of one's life history (Davidson 2001). As suggested by some researchers, this belief in the "undirected" nature of the sex drive cleared the way for a nonpathological view of homosexuality (79). However, Freud's view of the sexual drive (or at least the popularization of his view) simultaneously contributed to the further pathologization of sexual disinterest.

Interestingly, for all his focus on sexual drive and libido, Freud was not himself very interested in specific diagnostic categories for desire disorders per se. He developed only one specific diagnosis related to sexual disinterest: "oedipal inhibition of sexual love for a spouse" (Levine 2003, 283). However, in Freud's overall system, sexual disinterest can only ever be conceptualized as stemming from either a neurosis (specifically neurasthenia) or a psychic defense mechanism. "Neurasthenia" was a diagnosis developed in the 1880s, defined as fatigue from exhaustion of the nervous system, often thought to be the result of modernity. Freud thought neurasthenia could be caused by certain sexual activities, such as masturbation, when engaged in excessively, and that sexual neurasthenia could be a symptom of general neurasthenia (Laplanche and Pontalis 1973). In other words, in Freud's system, one explanation for sexual disinterest is that it is a symptom of a neurosis caused by excessive sexual activity. The second explanation is that the libido is in some way being blocked, denied, or transformed through a psychic defense mechanism, which are a part of both "normal" psychology and mental illness, and include repression, reaction formation, reversal into the opposite, and sublimation, among others (Laplanche and Pontalis 1973). Sublimation offers a somewhat positive explanation for sexual disinterest, as in that process, sexual energy can be redirected toward nonsexual aims, including socially valued activities, such as art and intellectual inquiry (Freud 2000; Laplanche and Pontalis 1973). However, the concept of sublimation still assumes that each person possesses a sexual drive, though some redirect part of this drive to nonsexual ends. In general, Freud's system asserts the universality of sexual desire—disinterest can only be the result of psychic defense mechanisms acting to block, deny, or transform the instinct. As we will see, in later years, psychoanalysts drew on Freud's work to create the diagnostic category of "inhibited sexual desire."

The 1930s to the 1960s: Narrowed Definitions of Frigidity and Professional Focus on Other Sexual "Disorders"

During this period, medical and mental health professionals remained interested in frigidity; however its diagnosis was narrowed to exclude sexual disinterest. This narrower definition was already taking shape at the turn of the twentieth century, reflected in the marriage manuals written for white, middle-class couples in the 1910s through the 1930s (Cryle and Moore 2012; Neuhaus 2000).[22] In these manuals, frigidity was defined as a woman's lack of "responsiveness" or pleasure during sex (McLaren 1999; Neuhaus 2000).[23] In the 1920s, the definition of frigidity was further narrowed in psychoanalytically inspired writing (Moore and Cryle 2010; Moore 2009). Freud himself claimed that, to reach psychosexual maturity, a woman must switch from clitoral to vaginal sexuality (Freud 1964; McLaren 1999). In the 1920s and 1930s, a number of his followers began to define frigidity specifically as the inability to experience orgasm through vaginal intercourse (Cryle and Moore 2012).[24] This definition survived until the 1970s and was accepted by many medical and mental health professionals in the United States in the post–World War II era (Cryle and Moore 2012; Lewis 2010).[25] US doctors in that era were concerned about white female frigidity and white male impotence, because they saw sexual satisfaction as the foundation of lasting marriages, which were seen as the foundation of American society. By threatening sexual satisfaction within marriage, female frigidity and male impotence threatened the stability of white heterosexual families and thus that of US power itself (Lewis 2010).

During this time, the central locus of sexology shifted from Europe to the United States (Irvine 2005). Influential American sexologists in this period, including Alfred Kinsey, William Masters, and Virginia Johnson, did not include specific diagnostic categories for sexual disinterest in their writing, but their research laid the groundwork for the (re)invention of a diagnostic category for hypoactive desire disorder in the 1970s. Kinsey (1894–1956) was primarily interested in the sexual behavior of white Americans, particularly that leading to orgasm. He was ostensibly uninterested in sexual desire, classifying certain types of sexuality as pathological, or developing treatments for sexual disorders.[26] However, because Kinsey saw sexual expression as natural and good, he was at best ambivalent about instances of nonsexuality. He did find evidence of it, particularly in women, and he created a separate category ("X") on his "heterosexual-homosexual" scale to describe these people.[27] However, in his description of those in the category, he wrote: "It is not impossible that further analysis of these individuals might show that they do sometimes respond to socio-sexual stimuli, but they are unresponsive and inexperienced as far as it is possible to determine by any ordinary means" (Kinsey et al. 1953, 472). Similarly, he noted that two percent of the women in his sample had never been aroused erotically; however, he again included a caveat: "It is, of course, not impossible that some of them had reacted erotically without being aware of the nature of their emotional responses; and it is possible that some other method of gathering data, or specific physiologic measurements, might have shown that some of these females had, on occasion, responded to erotic stimuli" (513). Thus, in the face of evidence of nonsexuality, Kinsey responded with the possibility that better scientific techniques may be able to ferret out the sexual response hidden beneath.[28]

Kinsey laid the foundation for William Masters and Virginia Johnson to conduct their research into the physiology of sex in the 1950s and 1960s. Using a variety of instruments, they measured research subjects' physiological changes when engaged in masturbation and partnered sexual intercourse in the laboratory.[29] In 1966, they published *Human Sexual Response*, which summarized their findings and proposed a four-phase model of the human sexual response cycle (excitement, plateau, orgasm, and resolution).[30] In addition, they identified so-called sexual dysfunctions that could occur at each stage of the cycle and developed a treatment program for these, all of which they described in their 1970 volume *Human Sexual Inadequacy* (Irvine 2005; Tiefer 2004).

Like Kinsey, Masters and Johnson were little interested in sexual desire. Their four-stage model of human sexual response did not include a desire/attraction stage. They did not focus on sexual disinterest as a disorder nor did they describe treatments for it in *Human Sexual Inadequacy* (Irvine

2005).[31] Leonore Tiefer (2004) speculated that Masters and Johnson may have omitted sexual drive or libido from their research, despite the importance of this concept in the history of sexology, because they were committed to achieving scientific legitimacy and respectability through the measurement of observable phenomena. According to her, sexual drive or libido is a variable phenomenon, difficult to observe and difficult to measure, and thus may have been unappealing to Masters and Johnson (42).

The Late Twentieth Century: The (Re)Invention of Hypoactive Desire Disorder in the United States

In the late twentieth century, the term *frigidity* fell out of favor, although it was replaced by terms like *orgasmic dysfunction* and *sexual arousal disorder* (Elliott 1985), conditions that have retained the interest of medical and mental health professionals up to the present.[32] While Masters and Johnson did not focus on sexual disinterest as a pathological condition, they were responsible for establishing the field of sex therapy, which quickly took up disinterest as a problem to be solved. According to Janice M. Irvine (2005), inhibited sexual desire (ISD), or hypoactive sexual desire (HSD), was created as a diagnostic category in the United States in the 1970s; in 1977, two sex therapists, Harold Leif and Helen Singer Kaplan, wrote independently about what they saw as the increasing prevalence of patients complaining of low libido, and called for its recognition as a distinct illness category.

Kaplan, a leading figure in modern sex therapy, reintroduced many psychoanalytic concepts and techniques. She refigured Masters and Johnson's "human sexual response cycle" as a three-step process involving desire, excitement, and orgasm (thus adding a desire stage, combining the excitement and plateau stages, and removing the resolution stage). She also employed psychoanalytic talk therapy to treat more "severe" cases of sexual dysfunction (Irvine 2005; Kaplan 1979).

Her most famous book, *The New Sex Therapy*, did not focus on desire disorders, but in 1977 she published an article about "hypoactive sexual desire" and in 1979 a book about the topic titled *Disorders of Sexual Desire and Other New Concepts and Techniques in Sex Therapy*, in which the case study at the beginning of this chapter appears (Kaplan 1977, 1979). In these publications, Kaplan argues that despite the high prevalence of "desire phase disorders" and the fact that they cause extreme distress to patients, inhibition of desire had been ignored by sex therapists.[33] She describes sexual desire as "an appetite that has its focus in the brain," with its own neural apparatus (1977, 3). Kaplan claims that desire can be inhibited by physical causes (including central nervous system depression, the use of drugs, and/or illness) or by psychic forces (such as conflict or fear). She states that both men and women experience hypoactive sexual desire although they may present differently; men with hypoactive desire often avoid sex because they fear not being able to perform, while women may adopt a passive role during intercourse (1977).

Furthermore, Kaplan states, patients with hypoactive sexual desire frequently deny that they are experiencing it: "There is a great deal of *denial* in this population. It seems easier to complain about orgastic and/ or erectile difficulties" (7). All sexual dysfunctions are ultimately caused by anxiety, she says; in her experience, the anxiety associated with hypoactive desire is typically more intense than that associated with excitement and orgasm phase disorders: "Patients exhibiting hypoactivity of sexual desire tend to be, as a group, more injured, more vulnerable, and, therefore, more rigidly defensive than those whose orgasm and/or erection are impaired while desire remains active" (8–9). She asserts that patients with hypoactive desire also fear pleasure and success in many other areas of their lives and are often difficult to treat. Although a few patients benefit from short-term interventions, most with hypoactive desire need long-term help. However, as we saw in the case of "Donald" and "Donna," she asserts the continuing usefulness of sexual therapy for hypoactive desire: "The process of sexual therapy is valuable because it will often confront the patient with the facts of his/her sexual and pleasure inhibition, with the fact that he/she does not allow himself/herself the full pleasure of making love.... This will sometimes motivate the patient to seek more profound and lengthy help, and this help must extend beyond the area of sexuality" (9). In *Disorders of Sexual Desire*, Kaplan does include a single paragraph in which she seems to recognize and legitimate asexuality as a sexual orientation, writing that "asexuality is certainly not always abnormal." According to her, for some people, "sexual appetite falls on the low side of the normal distribution on the basis of constitutional determinants," and they may not be bothered by this. She also states that a couple can learn to accommodate an "imbalance in sexual appetite," and an individual "can learn to accept and value himself [*sic*] apart from the strength of his [*sic*] sex drive" (1979, 68). However, given that this is a single paragraph in a book focused on the diagnosis and treatment of "disorders of sexual desire," this brief statement about "normal asexuality" is significantly undercut by its context.

Inhibited or hypoactive sexual desire was incorporated as a diagnostic category in the third edition of the American Psychiatric Association's (1981) *Diagnostic and Statistical Manual of Mental Disorders*. Neither of the first two editions of the *DSM* includes sexual disinterest as a mental illness. In the first, impotence, frigidity, and premature ejaculation are treated not as separate mental disorders but as possible symptoms of an underlying mental illness (American Psychiatric Association 1952). In the second, a diagnosis called "psychophysiologic genito-urinary disorder" was created and defined as "disturbances in menstruation and micturition, dyspareunia, and impotence in which emotional factors play a causative role" (American Psychiatric Association 1968).

Both Leif and Kaplan served on the "psychosexual disorders" advisory committee for the development of the *DSM-III*, in which "psychosexual disorders" is a separate metacategory, which included four broad categories: gender identity disorders, paraphilias (including homosexuality), psychosexual dysfunctions, and other psychosexual disorders. The first category included the subcategories of asexual, homosexual, heterosexual, and unspecified. Kaplan's reformulation of Masters and Johnson's sexual response cycle and the sexual disorders associated with each phase became the basis for the category of "psychosexual dysfunctions," in which the diagnosis for sexual desire disorder was called "inhibited sexual desire" (American Psychiatric Association 1981). In the revised third edition (*DSM-III-R*), the category "inhibited sexual desire" was renamed "sexual desire disorders" and divided into "hypoactive sexual desire disorder" and "sexual aversion disorder."[34] In addition, clinicians were asked to specify whether disorders were lifelong or acquired, universal or situational, and of psychogenic or mixed etiology (American Psychiatric Association 1987). The only major change made for the fourth edition (*DSM-IV*) was the addition of a "clinical significance criterion" to all of the sexual dysfunctions—in order for someone to be diagnosed with HSDD, the lack of sexual desire now had to cause "marked distress or interpersonal difficulty" (American Psychiatric Association 1994). The sexuality specification categories for "gender identity disorder" were changed to "sexually attracted to males," "sexually attracted to females," "sexually attracted to both," and "sexually attracted to neither" (Cohen-Kettenis and Pfäfflin 2009). No real changes were made to the 2000 text-revised version of the fourth edition (*DSM-IV-TR*).

In the 1980s, HSDD was considered a serious problem by sex therapists and other mental health professionals in the United States. According to a survey of 289 sex therapists published in 1986, 31 percent of patients sought therapy as a result of a "desire discrepancy between partners" and 28 percent did so as a result of individual desire problems (Kilmann et al. 1986). In addition, sexual desire problems were often considered by sex therapists to be extremely difficult to treat. As reported by some sex therapists, initially twice as many women as men sought therapy as a result of desire problems, although by the end of the decade, the gender discrepancy became less pronounced (Leiblum and Rosen 1988). Treatment included couples counseling, individual therapy, and sometimes medication (Irvine 1993). Most likely, the people who were able to access psychological counseling would have been predominantly middle- or upper-class white patients. Thus, as in the late nineteenth and early twentieth centuries, few people of color would have been diagnosed with a desire disorder (Irvine 2005). While in her 1977 article Kaplan suggested that homosexuality itself could be considered an example of situational hypoactive desire, in the 1980s, a number of sex therapists began to diagnose with HSDD and treat nonheterosexual individuals and couples who experienced comparatively low levels of same-sex desire. This was an especially important trend among lesbian-focused sex therapists and their clients, who were particularly concerned about perceived low levels of sexual activity in their relationships, a phenomenon referred to as "lesbian bed death" (M. Nichols 2004; Iasenza 2002; Hall 2004; Gupta 2013b; Przybylo 2019). Sex therapists also began to diagnose people with disabilities with HSDD and offer treatment specifically aimed at this population (Bullard 1988).

Irvine (2005) identifies a series of broad trends that contributed to the construction and popularization of HSDD in the 1980s and 1990s, one of which was the growth of sexology itself, as its practitioners sought professional recognition and a viable commercial market. Broader cultural shifts, such as those caused by the sexual revolution and the women's liberation movement, may also have contributed to the popularity of HSDD. As Irvine suggests, as a result of these social movements, more women began to expect and even demand sexual satisfaction. Yet, this demand was not met with the kind of structural changes that would have been required for women to exercise sexual autonomy, which may have led to sexual dissatisfaction (181).

Recent Developments Related to HSDD in the United States: Consolidation and Resistance

The recent history of the pathologization of sexual disinterest has been discussed by a number of researchers, including Leonore Tiefer (2006), who focuses on what she sees as the inappropriate medicalization of female sexual dissatisfaction by pharmaceutical companies; Thea Cacchioni (2015a), who focuses on the labor women are expected to perform in order to overcome their sexual "dysfunctions"; and Alyson K. Spurgas (2020), who focuses on the way in which medical and psychological efforts to address women's sexual dissatisfaction often overlook their experiences of gendered and sexual trauma. I analyze some aspects of this recent history through an asexual analytic in the fourth chapter of my book *Medical Entanglements: Rethinking Feminist Debates About Healthcare* (Gupta 2019b). Here I highlight key moments in the recent history of medical approaches to sexual disinterest.

On the one hand, during the past thirty years, there has been increased attention from doctors and researchers to sexual desire disorders, particularly among women, as well as to their physiological (as opposed to psychological) causes. The reasons for this are likely similar to those that led to a focus on "inhibited sexual desire" in the 1970s and 1980s, namely, economic considerations and the disconnect between people's expectations of sexuality and their actual sexual lives. In the 1980s and 1990s, doctors redefined "impotence" as a physiological problem (now called "erectile dysfunction," or ED), treatable by Viagra and other similar drugs (Marshall 2006). Allegedly as a result of the economic success of Viagra, pharmaceutical companies and associated researchers developed a new term for female sexual disorders (*female sexual dysfunction*, or FSD) and investigated various drug treatments for it. Viagra was tested for "female arousal disorder," but its effectiveness was limited. Arguably, as a result of their failure to find a drug treatment for "female arousal disorder," pharmaceutical companies turned their attention to female desire disorders (Tiefer 2006; Hartley 2006).

As will be discussed in the following chapter, researchers at this time had ostensibly moved beyond a "drive model" of sexual desire, instead adopting an "incentive motivation" model. However, they continued to see "sexual motivation" as necessary for the health and happiness of individuals and relationships, and thus continued to view a lack of sexual desire or motivation as a pathological condition in need of treatment. Researchers first tested the effectiveness of testosterone (under the name Intrinsa) as a treatment for female desire disorders, but because of limited effectiveness and side effects, it was approved for use in some parts of Europe but not in the United States (Kingsberg 2005; Guay 2005; "Testosterone" 2009; Hartley 2006). After the limited success of Intrinsa, the German pharmaceutical company Boehringer Ingelheim sought FDA approval in 2009 for a serotonin agonist/antagonist (then called flibanserin) as a treatment for female desire disorder. In part due to the opposition of feminist activists (discussed below), the FDA delayed its approval. In 2015, the company Sprout Pharmaceuticals finally obtained approval for flibanserin, renamed Addyi (Baid and Agarwal 2018). In 2019, despite significant criticism, the FDA approved a second drug treatment for female hypoactive sexual desire disorder, bremelanotide (Vyleesi), a melanocortin receptor agonist (Mintzes et al. 2021; Dhillon and Keam 2019).

Echoing the historical racialization of diagnostic categories for sexual disinterest, participants in clinical trials leading to FDA approval of Addyi and Vyleesi were overwhelmingly white (Matthew-Onabanjo et al. 2024). However, researchers have called for more attention to desire disorders among African American women (West et al. 2008), and studies using DSM diagnostic criteria for female sexual interest/arousal disorder and/or diagnostic instruments developed by US and European researchers, such as the Female Sexual Function Index and the Female Sexual Distress Scale, have been conducted in countries all over the world, including India, Iran, Brazil, and Ghana (Singh et al. 2020; Fooladi et al. 2020; Wolpe et al. 2017; Ibine et al. 2020), indicating that, as in other cases of gender and sexual pathologization, categories developed by American and European researchers with reference to white patients are often exported to the global majority.

While there has been significant attention paid to female desire disorders in the field of sexual health in the United States, there has also been resistance to the pathologization of sexual disinterest from both US feminist activists and ace online activists. A group of American feminist activists and academics, called "the New View Campaign," organized against the marketing of FSD as a disorder and lobbied the FDA to not approve flibanserin and similar drugs, arguing that the attempt to develop drug treatments for FSD is primarily profit motivated. In addition, it has suggested that the marketing of FSD pathologizes benign variation in levels of sexual desire; inappropriately "individualizes" and "biologizes" the issue of sexual dissatisfaction, discounting the relational, social, economic, and political factors that affect sexuality; and ignores differences between women based on race, class, culture, levels of sexual desire, sexual education, experiences of sexual pleasure, and so on (Working Group for a New View of Women's Sexual Problems 2002; Tiefer 2006).

Ace online activists have also challenged the medicalization of sexual disinterest. In the 2000s, the American Psychiatric Association went through the process of revising the *DSM*, which had not been substantively updated since the publication of the *DSM-IV* in 1994. Ace online activists sent a report to the working group responsible for revising the sections related to sexuality, outlining their critiques of the diagnostic category of "hypoactive sexual desire disorder" from an ace perspective. The report did not recommend eliminating HSDD entirely, recognizing that there may be some for whom low sexual desire is distressing and who might benefit from medical and/or psychiatric treatment. The primary suggestions of the report were to (1) include "attraction to neither males nor females [*sic*]" as a sexual orientation category; (2) require that the lack of sexual desire cause distress to the person and not only "interpersonal difficulties" for a diagnosis to be made; and (3) add a clause specifying that the diagnosis of HSDD only applies to people who do not consider themselves asexual (Brotto 2010b; Gupta 2013a; Hinderliter 2015).[35] The latter two recommendations were accepted by the working group and included in the *DSM-5*, which was released in 2013. In it, the category of HSDD was retained for men, and a separate category was created for women—female sexual interest/arousal disorder—supposedly to reflect the fact that it is "normal" for women to experience "responsive" but not "spontaneous" sexual desire (for a further discussion, see chapter 2).[36] For both diagnoses, the accompanying text indicates that a "desire discrepancy," in which a person has lower desire for sexual activity than their partner, is not sufficient for a diagnosis. In addition, for both diagnoses, the text indicates that if a person's low desire is explained by self-identification as "asexual," then a diagnosis should not be made (Brotto 2010a, 2010b; American Psychiatric Association 2013). Thus, asexual activism resulted in significant limitations being placed on the pathologization of sexual disinterest.[37]

However, despite feminist and ace activism, sexual disinterest remains pathologized. Some activists and scholars contend that maintaining any kind of diagnosis related to low or absent sexual desire in the *DSM* is inappropriate, even if the diagnosis does not apply to people who identify as asexual and/or who are not distressed by their level of sexual interest. For instance, Leslie Margolin (2023b) argues that low sexual desire should be treated like same-sex sexual desire—even if a person is distressed by their own (low or same-sex) desire, that in itself should not be considered pathological.[38] She also states that, as demonstrated by a case study included in the APA's *DSM-5: Clinical Case Studies* (2014), the existence of a diagnosis for hypoactive sexual desire means that people with low or absent sexual desire are always already suspected of pathology and must prove themselves exempt from the category due to their identity or lack of distress. In *Medical Entanglements*, I contend that while people should be able to access the medical interventions they consider necessary for their own well-being, and thus retaining such diagnoses may be a short-term practical necessity, as a society we should move away from a diagnosis-based approach to medicine as reflected in the *DSM*, since that inappropriately pathologizes some people while denying others access to desired medical interventions (Gupta 2019b).

In the late nineteenth and early twentieth centuries, sexologists like Krafft-Ebing and Hirschfeld developed categories, including sexual anesthesia and aneroticism, to describe a lack of sexual desire. In the 1970s, sex therapists like Kaplan developed categories, including inhibited sexual desire and hypoactive sexual desire disorder, to describe a lack of sexual desire. Although there are significant differences between the diagnoses developed in these two periods (e.g., Kaplan thought the ultimate root of ISD was anxiety, an idea not expressed by Krafft-Ebing or Hirschfeld), there are also substantial similarities. In general, most researchers involved in the pathologization of sexual disinterest understood desire as a biological and/or psychological drive functionally designed to promote reproduction, relationality, and/ or pleasure. Thus, they saw disinterest as a disorder with effects extending beyond the domain of the purely sexual, disturbing a person's ability to form social bonds and feel pleasure in other areas of their lives. In turn, this provided a justification for medical and mental health professionals to attempt to "fix" people who were uninterested in sex, through counseling or medication or both. While some may have benefited from these diagnoses and treatments, they reflect the fact that medical professionals were largely unable to conceptualize nonsexuality as a potentially fulfilling way of being-in-the-world. By promoting these ideas in their writing and clinical practice, medical professionals contributed to compulsory sexuality and to reducing the space available for the social acceptance of nonsexuality.

One might argue that the existence of diagnoses for sexual disinterest in the late nineteenth century and today indicates that there have always been people "suffering" from this disorder and care professionals looking to help them. But looking at the supposed "harms" thought to result from sexual disinterest casts doubt on this interpretation. Implicit in the early sexological research on the topic is a fear of "racial suicide" or a concern that white Americans and Europeans will not have enough children. This, however, was and is not a legitimate concern. Up until the 1990s, scientists and medical professionals claimed that sexual disinterest can harm a partner or a relationship but have since agreed that this is not an appropriate reason for diagnosing a person with a disorder. The members of a relationship may have wants that do not align (of which sex is just one example), but it is the mismatch, not the (absence of) wants that can cause harm. In addition, some of the relational harm could be avoided if people did not expect sex to be a standard part of intimate relationships in the first place. Scientists and medical professionals continue to assert that low sexual desire can harm an individual, by causing them distress and/or denying them experiences of pleasure. However, again, much of the distress (if it is experienced) is due to the stigma against asexuality. In addition, sex is merely one of many pleasures; people are varied in terms of the activities they derive pleasure from, and no one is considered deprived if they do not enjoy every activity that someone else does. Rather than demonstrating the inherent disorder of sexual disinterest, I argue that the long history of its pathologization reveals the persistent influence of compulsory sexuality.

Pathological categories for sexual disinterest are the places where scientific and medical ideas about sexuality most directly interface with, and potentially harm, those who identify as asexual and/or experience little or no sexual desire or attraction. In the remaining chapters of the book, I take a step back from "applied science" (e.g., clinical uses) to examine different areas of "basic" scientific research that inform and are informed by applied science. In the next chapter, I offer an original reading of neuroimaging studies on sexual arousal and desire and neurological models of it through the analytics of compulsory sexuality and asexualities. I demonstrate that this research relies on and reinforces essentialized understandings of sexual desire through a focus on it as an innate drive, and through a determination to draw boundaries between sexual desire and other emotional and motivational states. At the same time, rereading this research allows for a rethinking of the concepts of sexual desire, arousal, and pleasure.

CHAPTER 2

Sex in the Machine

The Neuroimaging of Sexual Desire

IN 2022, NEUROSCIENTIST, PSYCHOTHERAPIST, AND SEX THERAPIST NAN Wise published an article on the website *Psychology Today* titled "Learning How to Work the Sexual Brain-Body Connection," with the subtitle, "Here Are Some Hot Tips from Neuroscience for Better Sex." In it, she argues that "the ability to experience pleasures (in and out of the bedroom) is not a luxury but a necessity for a well-functioning brain and overall well-being." According to Wise, her neuroscience research located a "genital sensory cortex" in the brain that is activated by genital stimulation. To increase one's ability to experience pleasure, she recommends strengthening the neural pathways that connect the genitals with this sensory cortex through repeated stimulation. Wise goes on to claim that sex is good for the brain because genital stimulation increases blood flow to it, which is "generally a good thing" (Wise 2022).

While her recommendations seem close to tautological (essentially, to improve your ability to experience pleasure, practice experiencing pleasure), her research, and her own popularization of it, has circulated in the media for years. In 2011, her coresearcher Barry Komisaruk presented preliminary results from their fMRI studies on female orgasm at the Society for Neuroscience annual conference. The team released images, along with an arresting video showing the changes in activation that occurred in Wise's brain as she stimulated herself to orgasm (Sample 2011).[1] The release of the research led to a small media storm. "Forget Su Doku, Orgasms Are Better for the Brain" reads the headline of one article (Pavia 2013). "Sex on the Brain: What Turns Women On, Mapped Out" reads another (Geddes 2011). Science writer Kayt Sukel prolonged media interest when she published several articles sharing her own experience serving as a test subject for one of Komisaruk and Wise's studies, pithily describing her contribution as "coming for science" (Sukel 2011, 2012).[2] Wise published an article in *The Atlantic* about her experiences of orgasming while being scanned, which she describes as "donating orgasms to science" (Wise 2014), generating yet more interest in their research.

Clearly, the media have been repeatedly intrigued by the "sexy" nature of the story, the firsthand accounts of participants, and the audiovisual materials provided by the research team. And to be fair, Komisaruk and Wise's methods *are* interesting. For their studies, the research team created a plastic mask for each subject based on a mold taken of their face. After the subject put it on, the mask was bolted to the scanner. This helped to keep movement to a minimum as the subject stimulated herself (and, in one study, was stimulated by a partner) to orgasm. In their publication about their third study, Komisaruk and Wise assert that this was the first fMRI study of orgasm elicited by self- and partner-induced genital stimulation in women and was successful in "enabling identification of brain regions involved in orgasm" (Wise et al. 2017).

This is undoubtedly methodologically innovative research, and, as a number of commentators pointed out, scientists studying sexuality often overcome both a lack of funding and a lack of respect (Sukel 2011, 2013; Roach 2009). However, their studies also raise some serious red flags, specifically in regard to their purpose, their applicability to sexual experiences outside of the lab, and their representativeness. Komisaruk asserts that the purpose of the research was to understand how pleasure works in the brain. He also hoped that the research would be employed to develop methods for blocking pain and neurofeedback therapy for what he called the "pathological" condition of "anorgasmia" (Sample 2011).[3] During neurofeedback therapy, people view their own brain activity on a screen in "real time" and direct their bodies to act in different ways based on this information (Sample 2011). This later "therapeutic" purpose is reflected in Wise's 2022 article in which she attempts to translate their research into "some hot tips . . . for better sex." Thus, their research, or at least the popularization of it, as well as its translation into neurofeedback therapy, feeds into the pathologization of women's nonsexuality and also the pressure placed on women to "fix" their sexual dysfunction, as discussed in the previous chapter (see also Tiefer 2006; Cacchioni 2015a; Labuski 2015; Spurgas 2020). Arguably, Komisaruk and Wise's studies are but one part of a broader apparatus aimed at bringing women's sexuality into alignment with heterosexist norms.[4]

It is also uncertain whether the research actually reflects or is applicable to sexual experiences outside of the laboratory. For most people, the sensation of achieving orgasm while inside an MRI machine, whether through self-stimulation or with a partner, is probably distinct from reaching climax in "the real world." One intriguing aspect of Sukel's account is her discussion of her preparation for the study: for two weeks, she practiced the art of achieving orgasm with a bell fixed to her forehead, striving to master it without any head movement. Given that the conditions in the laboratory studies differ from most typical real-life experiences of orgasm—a problem with MRI-based studies of other experiences as well—it is reasonable to ask whether the neural processes during orgasm in an MRI machine mirror those in real-life situations. In simpler terms, we cannot simply presume that all orgasms are identical, even when studying the same people.

Regarding representativeness, it is also worth considering the extent to which the findings can be applied to a broader population. The research conducted by Komisaruk and Wise is not the first to raise concerns about generalizability when exploring the biological aspects of sexuality. For example, scholars expressed reservations about the participant sample used by William Masters and Virginia Johnson in their groundbreaking research on the physiological aspects of human sexuality in the 1960s. In short, Masters and Johnson exclusively studied people who could regularly achieve orgasm through masturbation and heterosexual intercourse (Tiefer 2004; Irvine 2005). Although their sample may not have been representative, their data were integrated into the *Diagnostic and Statistical Manual of Mental Disorders* and served as the prevailing model of healthy sexual response for many decades (Tiefer 2004).

According to Sukel, Komisaruk told her that only a few study volunteers were unable to stimulate themselves to orgasm in the MRI machine. This suggests a few things: First, that maybe a fair number of women are perfectly comfortable with, or even turned on by, loud noises, confined spaces, curious spectators, and immobilization devices. It also suggests that Komisaruk and Wise's volunteers may have shared a specific type of sexuality. For their third study, they report that two of fourteen volunteers were unable to achieve orgasm during the scanning session (two more datasets were discarded due to technical errors). The ten women who made it into the study all described themselves as being "highly orgasmic" (Wise et al. 2017). As we know, one of the research participants was Sukel (although it is not clear which of their three studies she participated in). As it turns out, another was also a science writer, Mara Altman, who also published an article about serving as a participant (based on her description, I believe she was in the first study) (Altman 2010). It is unclear whether Wise was included as a participant or just a "guinea pig" in any of the three studies. It is interesting to speculate whether science writers looking for a story and/or a sex researcher hoping for results would "skew" the study in any way. Overall, the small sample size and its possible unrepresentativeness are not themselves a problem; the only danger is if the sexuality of these women is established as the "norm" against which all other types are judged (like what happened with Masters and Johnson's research).[5]

Although Komisaruk and Wise's research focuses on orgasm specifically, the concerns I have raised here about their research—in terms of purpose, relevance, and representativeness—preview many of those I have about neuroimaging research on sexual "response."[6] These concerns are elaborated on in this chapter, which offers an original reading of neuroimaging studies on sexual arousal and desire and neurological models of sexuality through the analytics of compulsory sexuality and asexual possibilities. In recent years, neuroimaging research has captured the public's attention and helped to shape debates about what it means to be human, thus making it an important area to analyze for those interested in questions of justice. I argue that reading neuroimaging studies on sexual arousal and desire through the analytics of compulsory sexuality and asexual possibilities reveals the ways in which this research relies on and reinforces essentialized understandings of sexual desire, specifically through its focus on desire as an innate drive and its determination to draw boundaries between sexual desire and other emotional and motivational states. As discussed in the introduction to this book, essentialized understandings of sexual desire narrow the possibilities for sexual diversity and pluralism. At the same time, rereading this research through the analytics of compulsory sexuality and asexual possibilities demonstrates that sexual desire, arousal, and pleasure are not distinct categories, either in the scientific research or the material world it seeks to capture, thus clearing the way for more flexible, embodied, and socially situated understandings of desire and attraction. Before turning to this research, I first offer an overview of critical and feminist and queer engagements with neuroimaging research.

Critical Engagements with Neuroscience

The US Congress designated the 1990s as "the Decade of the Brain" (Jones and Mendell 1999). Since that time, neuroscience has continued to grow as an academic field, both in absolute and relative size, and has increasingly influenced society in the areas of health care, law, education, and consumer marketing, among others (Altimus et al. 2020; Rochon et al. 2019). In addition, neuroscience research has been taken up and popularized by the media, which often utilizes its findings to explain differences between groups of people—for instance, differences between men and women are explained based on findings about "male" and "female" brains (O'Connor et al. 2012, 221). As scholars have stated, "the production of brain-related knowledge is culturally important, carrying implications for how people see themselves as individuals and human beings" (220).

The growth of neuroimaging, especially functional magnetic resonance imaging (fMRI), has been an important part of the growth of neuroscience. fMRI experiments often work by requiring participants to engage in a cognitive task, such as a memory game, while the fMRI scanner detects changes in blood oxygenation levels in their brains. Neuroimaging research (and its interpretation and use) has been interrogated by analysts in a number of fields. Martha J. Farah (2014), arguing in support of neuroimaging research, offers four categories of critique: First, scholars have raised questions about the relationship between published brain images and actual brain activity, pointing out, for example, that significant interpretative work is involved in creating these images. Second, there is debate about the statistical analysis performed in some neuroimaging research, which can lead to (false) positive results.[7] Third, commentators challenge the influence of neuroimaging studies in society, suggesting that they can be overly appealing and/or convincing to the public, although Farah contests the evidence for these claims.

A fourth critique challenges assumptions in neuroimaging research, questioning the idea that specific mental processes occur in fixed brain regions (a theory called "localization"), as well as the subtraction method, which was employed in early studies. This method isolates brain activity related to a specific cognitive function by comparing two similar tasks. Brain activation from a simpler task (e.g., imagining a 3D shape) is subtracted from a more complex one (e.g., imagining a 3D shape rotating), leaving only the areas presumed responsible for the cognitive process of interest (e.g., mental rotation of 3D shapes). However, this method assumes that mental tasks are consistently performed through the same process and in the same brain region. Finally, commentators have questioned the use of reverse inferences, that is, researchers' assumption that, because a specific part of the brain has been associated with a particular psychological process, anytime that part is activated, that process must be occurring (Farah 2014).

Farah (2014) and other scholars agree that while criticisms of localization and the subtraction method have some validity, these assumptions and methods were more prevalent during the early days of neuroimaging research. Jessey Wright (2018) argues, for instance, that recent neuroimaging research employs more complex methods and seeks more often to identify the distributed patterns of brain activity involved in mental processes. Still, even more recent neuroimaging studies should be analyzed carefully to ensure that they do not replicate flawed assumptions, methods, or forms of analysis.

In addition to the concerns outlined by Farah, analysts have also identified and challenged the "neurocentrism" underlying both neuroscience research and its mainstream popularization. Under that paradigm, the brain is solely responsible for all human thinking, feeling, and deciding. As critics note, this view disregards the rest of the body and reinforces Cartesian mind/body dualism (Halberg 2022).

They have also pointed out that, as in many areas of psychological research, neuroimaging studies have too often relied on small sample sizes. In 2010, psychologists pointed out that many studies make universal claims about human psychology based entirely on research performed in the Global North, often on undergraduate students. Yet there is evidence of significant psychological differences between even college- vs. noncollege-educated people from the Global North (Henrich et al. 2010). Feminist, antiracist, and postcolonial scholars have long criticized the equation of the white man with the "human" and the "universal" (Wynter 2003; Beauvoir 2010), but contemporary neuroimaging research continues to replicate this false universalization. While some neuroscientists have called for their field to think critically about sample choice and to carry out more neuroimaging studies with large, demographically diverse samples (Paus 2010; Falk et al. 2013), as we will see, most such research on sexual arousal and desire falls short of these goals.

Feminist and Queer Engagements with Neuroscience Research on Sex, Gender, and Sexuality

Over the past two decades, neuroscience has produced a growing body of research on the neurological aspects of sex/gender differences, sexual orientation, and gender identity.[8] Much of this body of research is at least partially informed by evolutionary and brain organization theories, described in this book's introduction. In short, under this paradigm, men and women have evolved different types of sexuality, and these are hardwired into the brain during fetal development as a result of hormonal action. "Atypical" sexual and gender identities are assumed to result from "atypical" prenatal hormone exposures, which are then reflected in "atypical" adult brain structure and function (Buss 1998; Okami and Shackelford 2001; Bakker 2019; Roselli 2018). Based on these background assumptions, neuroimaging studies have investigated differences in brain structure and functioning between male and female, heterosexual and homosexual, and cisgender and transgender subjects (Gaillard et al. 2021; Frigerio et al. 2021). In *Brain Storm: The Flaws in the Science of Sex Differences*, Rebecca Jordan-Young offers a comprehensive critique of the assumptions, methods, and analysis employed by brain organization researchers.

More generally, neurofeminist analysis has challenged many of the background assumptions underlying this research on gender, sexuality, and the brain, as well as the specific studies themselves. Most fundamentally, queer and feminist scholars have criticized neuroscience research that accepts the binary sex/gender system as a natural fact and have challenged its use to support gender inequality in broader society (Rippon et al. 2014; Schmitz and Hoppner 2014).[9] A 2014 review by neurofeminist scholars proposed four guiding principles for sex/gender neuroimaging research: recognizing overlap between men and women, acknowledging sex/gender mosaicism (that most people combine "male" and "female" characteristics), considering how gender differences vary across contexts (here they also urge scientists to consider intersectionality), and examining how social norms shape brain structure and function. After outlining these broad principles, the authors offer a series of concrete suggestions to improve sex/gender neuroimaging research design, analysis, and interpretation, such as measuring sex/gender in a multidimensional way and using this research to assess changes in sex/gender differences over time or across contexts (Rippon et al. 2014; see also Friedrichs and Kellmeyer 2022).[10]

Expanding on this earlier call for intersectionality, Annie Duchesne and Anelis Kaiser Trujillo (2021) identified an ongoing absence of intersectionality in neuroscience research more broadly, in neuroscience research on sex and gender specifically, and in neurofeminist analyses of neuroscience research. They argue for the need for intersectional research in neuroscience and in feminist scholarship on neuroscience (neurofeminism), writing, "Incorporating approaches from intersectionality can inform the study of [social] categories while promoting research that measures or otherwise accounts for their interdependency rather than falsely orthogonalizing them" (10).[11]

Queer and feminist scholars have also analyzed neuroscience research on sexuality, mostly focusing on neuroimaging studies of sexual orientation. For instance, Isabelle Dussauge and Anelis Kaiser (2012) reviewed research conducted between 2000 and 2010.[12] They found that the studies employed a 2 × 2 framework, dividing participants into binary gender categories (male and female) and binary sexual orientation categories (heterosexual and homosexual). They also explicitly or implicitly assumed that both gender and sexual orientation were fixed for the participants. Other aspects of sexuality were ignored. Dussauge and Kaiser encouraged neuroscientists to approach sexuality as multidimensional (i.e., more than just the gender(s) a person is attracted to), multiple (i.e., more than just heterosexual vs. homosexual), and potentially fluid over time and in different contexts. Commentators also critique the neuroscience research on sexuality for continuing to link homosexuality and gender inversion (the idea, for example, that gay men are "like" women)—a theory that has its roots in nineteenth-century sexology (Jordan-Young 2010). More recently, others have asserted that neuroscience studies on (trans)gender identity similarly rely on essentialized understandings of sex/gender, ignoring socially situated, developmental, mosaicist, and depathologizing approaches (Llaveria Caselles 2021; Wang 2022).

Neuroimaging Research on Sexual "Response"

For this chapter, I reviewed over thirty English-language articles published between 1990 and 2021 on neurological models of sexual desire and/or neuroimaging studies of sexual response, arousal, or desire. My sample included articles reporting the results of individual neuroimaging studies (e.g., Mouras et al. 2003; Ortigue and Bianchi-Demicheli 2008), theoretical or opinion articles (e.g., Stoléru 2014; Gola et al. 2016), qualitative reviews of research (e.g., Diamond and Dickenson 2012; Cheng et al. 2015), and quantitative meta-analyses of research (e.g., Poeppl et al. 2016; Kühn and Gallinat 2011). I intentionally included highly cited articles and those by scientists with multiple publications in the field. Of those that reported on original research studies, almost half were conducted by researchers in Germany. The remainder were conducted in Russia, the Netherlands, France, the United States, South Korea, or China. In my review, I analyzed the articles with the following questions in mind: How are scientists defining and operationalizing different aspects of sexuality, particularly concepts like sexual response, arousal, and desire? What samples are being used? Are they considering intersubject diversity or intrasubject change over time? How are social categories like gender, race, and sexual orientation operating? Based on this review, I argue that at least a significant part of fMRI research on sexual response relies on and reinforces an essentialized understanding of sexuality as a unique, innate drive. Alternatively, other neuroscience research on sexuality allows for a nonessentialized understanding of sexuality, and this, combined with a developmental systems approach, can offer a flexible and ace-positive model for understanding sexual attraction and desire.

Background Assumptions: Models of Sexual Desire,
Arousal, and Response, Part I

Here I briefly review the main background assumptions about sexuality and models of sexual desire utilized in the examined research. After looking at the neuroimaging studies themselves, I offer a more detailed critique of the models of sexual desire operating in such research on sexual desire, arousal, and response.

A number of scientists recognize that sexual desire remains a "slippery" concept (Brotto and Smith 2014). For example, James G. Pfaus (2009, 1506) begins his review article titled "Pathways of Sexual Desire" with the following statement: "Sexual desire seems a straightforward concept, yet there is no agreed-upon definition of what it is or how it manifests itself." Researchers point out numerous limitations of the concept, acknowledging that there is no consensus on the difference between sexual desire and sexual arousal, or whether one precedes the other. Similarly, some believe there exists a difference between spontaneous (or intrinsic) and responsive sexual desire, while others dispute this claim (Brotto and Smith 2014; Meana 2010). Some, but not all, have claimed that there is a distinction between desire for solitary sexual activity and that for partnered sexual activity (Spector et al. 1996). Scientists also acknowledge that while there has been a dearth of social scientific studies investigating how people define and experience desire, the existing research suggests that desire can be an idiosyncratic experience. Reporting on a qualitative study of sexual desire with forty heterosexual women in committed relationships, Denisa Goldhammer and Marita McCabe (2011) note that some participants described their experiences of sexual desire as physical sensations; some as cognitive processes involving fantasies, memories, or conscious wants; some as emotional experiences; and some as interpersonal reactions, either triggered by a partner or as a force directed toward a partner. Many women described sexual desire as a complex interplay of these different elements. Social scientific research on men's subjective experience of sexual desire is even more limited, but small studies suggest that they may describe sexual desire as having genital, nongenital-physical, and cognitive-affective components (Dang et al. 2017). In addition, men may not consistently distinguish between desire and arousal, may experience desire and arousal sequentially (in either order) or simultaneously, and may describe one or the other as more mental, more physical, or some combination (K. Mitchell et al. 2014).

An examination of contemporary neurological models of sexual desire, arousal, and/or response reveals both an official rejection of the "drive" model of sexual desire and its lingering influence. Scientists in the field identify Freud as the most influential scientific proponent of the drive model, arguing that he saw all human beings as motivated by the pleasure drive, or libido. According to their understanding of his theory of libido, Freud conceptualized sexual desire as an innate system that leads to the buildup of pressure or energy in people, which motivates them to pursue certain activities to relieve this pressure, and this relief is experienced as pleasure (Both et al. 2007; Toates 2009). Although Freud saw the libido as flexible in both its aims and objects, this essentialized model of sexual desire is limited in its ability to understand asexuality as nonpathological, claiming that it can only be understood as either a lack of a (or the) primary drive or its repression or sublimation (see chapter 1 for a further discussion).

In their historical narratives, the scientists in the field then note that sex researchers William Masters and Virginia Johnson, focused as they were on documenting physiological changes during sexual activity, left out the concept of desire altogether in their influential four-phase model of "human sexual response." In their histories, sex therapist Helen Singer Kaplan (and, to a lesser extent, Harold Leif) reintroduced the concept in the 1970s, adding "sexual desire" to the beginning of Masters and Johnson's model of human sexual response. Kaplan's conceptualization of this desire closely adhered to Freud's understanding of libido. As scientists in the field note, Kaplan's modified version of the human sexual response model (desire, arousal, orgasm) was incorporated into the *DSM* and its definition of hypoactive sexual desire disorder (Both et al. 2007; Stoléru 2014; Toates 2009) (again, see chapter 1 for a critical discussion of this history).

In their own narratives, many psychologists and neuroscientists in the field of sex research left behind the Freudian drive model of sexual desire in the 1980s and 1990s in favor of an incentive motivation model (Both et al. 2007; Stoléru 2014; Toates 2009). The latter arose out of a general incentive model of motivation developed in biological psychology in the 1970s and 1980s (Kringelbach and Berridge 2016; Toates 2009), which, applied to the realm of sexuality, views sexual motivation as "the result of the activation of a sensitive sexual response system by sexually competent stimuli that are present in the environment" (Both et al. 2007, 329).

Sex (?) in the Machine: Why?

Brain scanning technologies have been used by scientists to investigate different aspects of what is often called "sexual response" since the 1990s (e.g., Rauch et al. 1999). The stated purpose of most of these studies falls into one of three overlapping categories, which I call pathology, difference, and normalcy. In the category of pathology, studies attempt to contrast the brain sexual response of sexually "healthy" subjects to that of sexually "pathological" subjects, with the goal of developing treatments for the latter. Differences in brain sexual response have been explored through neuroimaging technology for a variety of pathologies, from hypoactive desire disorder (e.g., Bianchi-Demicheli et al. 2011) to hypersexual disorder and sex addiction (e.g., Seok and Sohn 2015) to erectile dysfunction (e.g., Cera et al. 2014) to pedophilia (e.g., Renaud et al. 2011). It is not surprising that pathology motivates a significant portion of neuroimaging research on sexuality; after all, scientific funding is often easier to secure when researchers can claim a practical application. Pharmaceutical companies have funded some of these studies; for example, Pfizer Pharmaceuticals has funded at least one on brain sexual response for "normal" women vs. those with hypoactive desire disorder (Arnow et al. 2009).

As discussed extensively regarding hypoactive desire disorder in chapter 1, many of these so-called pathologies are highly contested diagnoses.[13] The neuroimaging studies in the category of pathology generally rely without question on *DSM* definitions of these conditions. Articles about these studies often include statements about the (widespread) prevalence of these "disorders" as justification for the research and use the results to speculate about potential neuropharmacological treatments. These choices simplify and essentialize what are often complex biopsychosocial-*political* experiences of distress into self-evident disorders in need of research and treatment. It is worth noting that a small minority of these studies—specifically those on pedophilia—are also designed to produce forensic applications. These scientists want to develop ways to employ brain scanning technology to identify men with pedophilic interests, even when they are attempting to hide them (Ponseti et al. 2012; Wernicke et al. 2017). Thus, as in the heyday of nineteenth-century sexology, sex research remains connected to highly racialized carceral systems.[14]

In the category of difference, studies attempt to compare the brain sexual response of different groups of people (without so-called sexual pathologies). The most common comparison is between men and women, one that has been investigated by numerous neuroimaging studies (e.g., Strahler et al. 2018; Wehrum et al. 2013; Parada et al. 2016; Stark et al. 2019) and several quantitative meta-analytic reviews (e.g., Poeppl et al. 2016; Mitricheva et al. 2019), with contradictory results. The most recent meta-analysis (Mitricheva et al. 2019) found no sex/gender differences in brain sexual response, but the debate is far from over (Poeppl et al. 2020), and scientists continue to conduct these types of experiments (e.g., Putkinen et al. 2022). Other studies in the category of difference have compared the brain sexual response of heterosexual and homosexual subjects (e.g., Kagerer et al. 2011; Sylva et al. 2013), nonintersex and intersex subjects (Hamann et al. 2014), cisgender and transgender subjects (e.g., Gizewski et al. 2009), asexual and allosexual subjects (Prause and Harenski 2014), and women at different stages of the menstrual cycle (e.g., Abler et al. 2013).

These "difference" studies are subject to the same critiques that feminist and queer scholars have made of neuroscience and neuroimaging studies of sex, gender, and sexuality more generally, which I outlined at the beginning of this chapter. As noted earlier, many of these studies set out to investigate the validity of sexual selection theory, which posits the existence of evolved sex/gender differences in sexuality, and the brain organization hypothesis, which explains "typical" and "atypical" gender and sexual expression as the result of different prenatal hormone exposures. In her analysis of neuroimaging studies on sex/gender differences in sexuality, Alyson K. Spurgas (2020, 82) writes, "What can be gleaned from these studies is questionable at best, as many unstated narratives about sex differences in cognition, sexuality, and desire abound and inform hypotheses and experimental design—before the research has even begun." Spurgas also argues that, in recent years, neuroimaging research on sex/gender differences in sexuality has been incorporated into a broader effort by psychologists and sex therapists to define female sexuality as more "responsive" and "receptive" than male sexuality, to distinguish between "normal" and sexually "dysfunctional" women, and to establish the "disconnect" between subjectively experienced feelings of desire and objectively measured physiological arousal for many female test subjects (2012; 2020).

As also noted earlier in the chapter, feminist and queer critics of neuroimaging research on sex, gender, and sexuality have noted that few of these studies consider intersectionality, or the idea that systems of oppression (like racism, sexism, heterosexism, etc.) intersect. This holds true for neuroimaging research on sexual response also. The studies I reviewed were either single sex/gender studies or reported differences by sex/gender, and a few reported differences by sex/gender and sexual orientation, but, as noted above, studies tended to assume that sex/gender or sexual orientation differences were the result of biology rather than systems of gender or sexual inequality. Only a single study reported data by race or ethnicity (P. Chen et al. 2020). None examined the influence of racism or colonialism on sexuality, either alone or in combination with other systems of oppression.[15]

In the category of normalcy, studies seek to understand the unique aspects of the brain's "sexual response system," without explicitly focusing on pathologies or explicitly comparing different groups of people (e.g., Arnow et al. 2002). These studies all use heterosexual, sexually "healthy" subjects (often young adults) in order to understand the (one universal) brain's unique sexual response system.[16] Thus, results obtained from a socially "normal" research population are abstracted and universalized. In her analysis of neuroimaging studies of sexual arousal, sexual orientation, and orgasm, Isabelle Dussauge (2013) focuses on this effort by scientists to produce an abstract and idealized model of sexuality. She contends that the attempt to identify a single model of brain sexual arousal or response "reveals an ontological assumption: that there is a universal desire and pleasure which, once triggered, is the same for everyone" (131).

It is worth noting that, in fact, results from studies with only "normal" *male* research subjects have often been abstracted as the (one universal) brain's sexual response model. Most of the early neuroimaging studies of sexual response in the brain focused on male participants only (e.g., Mouras et al. 2003; Stoléru et al. 2003; Rauch et al. 1999; Arnow et al. 2002). A meta-analysis published in 2019 found that around 65 percent of research subjects in neuroimaging studies on sexual arousal up to that point were male, and around 89 percent of those were heterosexual (Mitricheva et al. 2019; see also van't Hof and Cera 2021). This created a large bank of information about brain sexual response based entirely on male minds and bodies, and thus set those as a norm. According to Sophie R. van't Hof and Nicoletta Cera (2021), one influential model of brain response to sexual stimuli is based primarily on data drawn from male subjects. In fact, the "dual control model" of human sexuality (or the idea that human sexual response involves both excitation and inhibition) was developed by researchers focusing on male erectile dysfunction (Pfaus 2009; Janssen and Bancroft 2007; Bancroft et al. 2009), and an influential four-component model of brain sexual response (cognitive, motivational, emotional, and autonomic) was developed based on neuroimaging research with male subjects (Stoléru et al. 1999, 2012).[17]

As noted above, almost half of the studies reviewed were conducted in Germany, and only three were conducted outside of the United States or Europe. As only one reported data by race or ethnicity, definite conclusions cannot be made, but it seems likely that most subjects were white. Thus, as in the past, data from white and/or male subjects are abstracted into a universal model, with its maleness and whiteness unmarked.

In addition to forming the basis for abstract models, the data from normalcy studies can be, and often is, used comparatively by scientists explicitly investigating sexual pathologies and nonnormative sexualities. As noted above, all of the normalcy studies I reviewed were either single-sex/gender studies (e.g., only men or only women) or reported results by sex/gender. Thus, they could be, and often were, employed in later qualitative reviews or quantitative meta-analyses explicitly investigating sex/gender differences in brain sexual response.

In contrast to most neuroimaging research on sexual response, some researchers have investigated how the entanglement of culture and society affects the processing of sexual stimuli, with much richer and more varied results. For instance, Mina Cikara et al. (2010) measured brain activation in response to sexualized images in a group of heterosexual men and then correlated imaging results with the participants' sexist attitudes toward women. They found that higher hostile-sexism scores for male participants predicted less spontaneous activation in response to sexualized images of the brain network associated with mentalizing. While there are several critiques that could be made of this study, it is notable for its effort to utilize neuroimaging research to address questions related not only to biological sex/gender but also sexist attitudes and sexism as a system of oppression.[18]

Sex (?) in the Machine: How?

As discussed at the beginning of this chapter, a few studies have employed neuroimaging technology while subjects have stimulated themselves, or been stimulated by a partner, to orgasm (e.g., Stoléru et al. 2012; Georgiadis 2011; Wise et al. 2017). However, most neuroimaging studies investigate sexuality by scanning participants' brains while they are exposed to sexual "stimuli," mostly using what scientists call "visual sexual stimuli" or vss (sexual photographs or videos).[19] The content of the vss selected varies, usually consisting of pictures or videos of naked people and/or people engaged in normative sex acts (Stoléru et al. 2012; pictures of genitals used in Ponseti et al. 2006). Most studies do not provide substantive detail about the erotic images or videos selected, for example, whether they include people of different body types, races, or ethnicities, but they likely feature people who meet hegemonic standards of sexual attractiveness. Most studies only measure response to sexual stimuli thought to be congruent with a research participant's sexual orientation (called "preferred sexual stimuli"); for instance, lesbian women are only shown erotic pictures of women.[20] A few imaging studies have also measured physiological arousal (such as blood flow to the genitals) while brain imaging was taking place (e.g., Moulier et al. 2006). Most also assess the participants' subjective sexual arousal or desire, usually through questionnaires (Stoléru et al. 2012). Studies that do not assess subjective sexual arousal or desire must assume that because others have found the stimuli sexually arousing, participants in the study will also.

What Is the Activity of Interest Here? Or, Do Brains Viewing Porn Equal Brains Doing Sex? Many of the neuroimaging studies reviewed state that they are investigating brain activation during sexual "response," arousal, and/or desire. But what they are actually investigating is brain activation during one very specific "sexual" activity: passively viewing normatively erotic images or videos. In her article "Between the Screens: Brain Imaging, Pornography, and Sex Research," Anna E. Ward (2018) comments on this positioning of the body between two forms of visualization: of film pornography and of the brain image. As Ward notes, this renders the body "more transponder than flesh and blood" (17). Brain activation during this one (potentially) sexual activity is generalized and universalized as "the brain's sexual response." Few studies compare brain activation from different types of sexual stimuli (e.g., photos, videos, pheromones), as long as they are, in general, "preferred" sexual stimuli.[21] Dussauge (2013, 138) contends that, as reported by researchers, ideal desire and pleasure are "triggered by specific situations but not qualitatively influenced by those."[22] In other words, they assume that brain activation will be roughly the same whether the participants are looking at pictures of naked people, watching pornographic videos, smelling human pheromones, fantasizing about sex, or listening to erotic stories.

In addition, the lack of qualitative data collection makes it difficult to attribute meaning to the brain activation patterns of participants looking at VSS. Beyond assessing levels of arousal or desire, none of the studies I reviewed provided any more information about the subjects' experience of the research tasks. Were some approaching them as "foreplay"? Were some experiencing them as a sexual activity itself? Did the participants experience looking at erotic photos in a lab while their brains were being scanned as different from looking at such pictures in other contexts? Should the results be interpreted even more narrowly—not even as brain activation while viewing erotic pictures or videos but as brain activation while viewing normatively erotic pictures or videos while under observation by scientists for the purpose of knowledge creation?

A few scientists have raised these types of questions. Several note that in neuroimaging studies, subjects are instructed not to move and therefore may be inhibiting their responses to sexual stimuli in a way they might not outside the lab (e.g., Fonteille and Stoléru 2011; Stoléru et al. 2012; Gola et al. 2016). In a commentary, Gola et al. (2016) argue for the importance of distinguishing whether vss act as "conditioned stimuli" (cues predicting an upcoming reward) or "unconditioned stimuli" (rewarding by themselves). Reviewing forty neuroimaging studies using vss, they found that nine described vss as cues, sixteen as rewards, and one as both; fifteen did not utilize any such labels. They claim that it is important to conceptualize vss as cues or rewards in order to interpret neuroimaging results.[23] For my purposes, it is not important to adjudicate whether vss are cues or rewards, but to point out that it is not necessarily obvious what is going on (sex? anticipation of sex?) when participants view vss in a lab.

The use of the term *sexual response* reflects the ambiguity of what is actually going on in these studies. In other contexts, sex researchers employ terms such as *sexual arousal*, *sexual desire* or *motivation*, *sexual pleasure*, and/or *orgasm*. In neuroimaging studies, they use the term *sexual response* because that is literally what they are measuring (how the brain responds to sexual stimuli), but they tend to assume that the term is in some way reflective of sexual arousal, desire, (less explicitly) pleasure, or some combination of these, but they do not collect the qualitative data that would be required to make these connections in a nuanced way.

What Is Being Compared? Or, Is There Something Unique About Sex? Most studies of sexual response, arousal, and/or desire employ the subtraction method, mentioned at the beginning of this chapter, in which brain activation in response to a "nonsexual" stimulus is subtracted from brain activation in response to a sexual stimulus, sometimes combined with additional analysis.[24] In theory, the nonsexual and sexual stimuli should only differ in one respect—nonsexual vs. sexual—to allow scientists to isolate those areas of brain activation that are unique to sexual response. Of course, this protocol assumes that certain images or videos are not "sexual" and will not be experienced as such by research subjects. As Dussauge (2013, 133) notes, "What counts as sexuality is thus defined as much by what does not count as sexual pleasure/desire."

In their own background assumptions, researchers conceptualize "sexual" response as most similar to other emotional and motivational states. As discussed earlier, many adopt an incentive motivation model of sexual response, which views sexual motivation as similar to other motivational states or action tendencies, including hunger, thirst, and drug-craving (Toates 2009). Some scientists, notably Morten L. Kringelbach and Kent Berridge, argue that there exists a common brain "reward" or "pleasure" system for all types of rewards or pleasures, which includes three component processes: "wanting" (understood as desire or motivation), "liking" (understood as pleasure), and learning (2016; Georgiadis and Kringelbach 2012). The studies reviewed for this chapter made frequent references to incentive and/or reward motivation models, and often used studies on other emotional or motivational states to guide their own research and interpret their own results. However, most neuroimaging studies utilize what are described as "neutral" photos or videos as the comparison/subtraction task; only a minority directly compare sexual response to other emotional or motivational states, and, to my knowledge, none directly compares sexual response to any of the other three most commonly cited motivational states—hunger/eating, thirst/drinking, and drug-craving/drug-taking—although some of these comparisons have been made via meta-analysis.[25] Dussauge (2013, 133) reasons that in employing the subtraction method to isolate the specifically "sexual" aspect of brain sexual response, researchers are addressing and containing what she calls an "epistemological anxiety of specificity," stating: "The epistemologically threatening question is this: what in the sexual response makes it, specifically, sexual and not just emotional (cf. Walter et al. 2008)? What is the brain activation pattern which makes an orgasm, an orgasm—and not, say, simply a pleasurable bodily experience?" I agree with Dussauge that scientists are committed to isolating "the" sexual response system as universal and unique.

However, given how often they select inadequate comparison tasks, I argue that they also avoid truly confronting what Dussauge calls "the anxiety of specificity," which I also call "a fear of undifferentiated desires and pleasures."[26]

The type of nonsexual stimuli used varies across studies, although many do not provide details. For those measuring orgasm, the comparison task was usually a resting state. For ones measuring response to VSS, the comparison task was often described as "neutral" pictures or videos, such as those of sports, nature, "nonsexual" interactions between people, household objects, and/or "mosaic images" (Stoléru et al. 2012). However, it is likely that there are many differences between erotic photos and "neutral" ones beyond the sexual vs. nonsexual difference. Depending on what is selected for the "control" stimuli, the variation in brain response could be attributed to, for example, the difference between viewing people and objects, between emotionally intense vs. emotionally mild stimuli, and/or between stimuli that provoke a desire to move vs. those that do not. Thus, it is overreaching to claim that the studies that employ "neutral" stimuli identify a specifically "sexual" brain response system.[27]

Many researchers are aware of the limitations of selecting neutral photos or videos, which makes the prevalence of the practice perhaps more surprising. For instance, in the discussion section of an early study, Jérôme Redouté et al. (2000, 174) note, "Our results support the notion that SA [sexual arousal] is a composite psychophysiological state correlated with the activation/ deactivation of several brain regions. Among those regions, a majority . . . have been associated with other emotional states." They recommend that future studies should employ control stimuli that generate the same level of emotional arousal as the sexual stimuli. They also note that their own control stimuli (humorous videos) did not generate a motivational state in the way that their sexual stimuli did, and recommend that scientists should select control conditions that generate motivational states such as "anger or greed for savory foods" (175). Despite recognizing the limitations of their own study, they remain committed to the idea that sexual response is distinct from other brain activation patterns, writing that the "specificity" of sexual response "may be related to: (i) a distinctive pattern of activated/deactivated areas and/ or (ii) the activation/deactivation of discrete areas within the broad regions demonstrated by PET" (174). Yet, two decades later, only some of the studies they suggested have been carried out.

Some studies have compared sexual response to other emotional states, but often the comparisons still lack a certain robustness. A few have compared sexual response to a disgust response, as both are thought to produce high levels of emotional arousal, and research has indicated that people can experience mixed emotions, including desire and disgust, in response to erotic material (for a discussion, see Peterson and Janssen 2007).[28] For example, a study by Charmaine Borg, Peter J. de Jong, and Janniko R. Georgiadis (2014) focused on the co-occurrence of sexual and disgust responses and found considerable overlapping brain activity when viewing pictures of sexual penetration and "nonsexual" disgust-inducing ones (e.g., a person vomiting, feces), as well as a correlation between a subject's response to vss and their implicit association of disgust and sexual penetration, indicating that at least some brain activation in response to vss may reflect disgust. While this study is useful for its recognition of the co-occurrence of different emotions in response to vss, it only included women and was part of a broader research agenda focused on treating "sexual dysfunctions" among women (see also Borg et al. 2014).

To my knowledge, the first neuroimaging study to attempt to systematically isolate sexual response from all other emotional responses is a study by Martin Walter et al., from 2008, almost a decade after the publication of the first neuroimaging studies of sexual response.[29] In it, participants were shown five types of images (positive erotic, negative erotic, positive emotional, negative emotional, and neutral) in an attempt to distinguish between general emotional effects and specifically sexual ones. Little information is provided about the emotional pictures, which are described as depicting "one or more humans in sports scenes or other social interactions as well as emotionally arousing non-human motives in a positive or negative context" (1483).[30] As suggested by Redouté et al. (2000), the researchers did choose sexual and emotional images that have been previously rated similarly in terms of arousal and valence, but the actual participants did not agree with these ratings for most of the pictures.[31] Overall, the study found significant overlapping areas of activation for sexual response compared to other emotional responses and some unique areas of activation.

A 2013 study by Sina Wehrum et al. compared brain response to four types of stimuli: sexual, neutral, positive, and negative. In it, the "positive" category included "nonsexual scenes" such as "sport/adventure scenes" and "people in funfairs." Sophie R. van't Hof et al. (2022) note that while Walter et al. (2008) and Wehrum et al. (2013) identify some common areas as specific to "sexual" response, their findings differ as well, despite both attempting to isolate sexual response from emotional response. Employing tools of machine learning, van't Hof et al. (2022) use data from Wehrum et al. (2013) and Kragel et al. (2019) to construct a computer model that can accurately categorize a brain scan as reflecting activation in response to either sexual or affective images, at least among young, heterosexual, cisgender, sexually "healthy" adults from Germany or Boulder, Colorado.[32]

Another comparison that has been investigated is the difference between brain response to sexually preferred vs. nonpreferred stimuli (Sylva et al. 2013; Safron et al. 2007; Paul et al. 2008; Hu et al. 2008; Zhang et al. 2011). As noted earlier, most studies only show participants stimuli congruent with their sexual orientation (preferred), but some show incongruent stimuli, for instance, gay men may be shown erotic pictures of men and others of women.[33] What is interesting about this set of studies is the fact that it is not always clear how scientists are conceptualizing or participants are experiencing the nonpreferred stimuli—as "sexual but less sexual" than preferred stimuli, as neutral, or as aversive. Sylva et al. (2013) specifically exclude photos with high disgust ratings, thus the nonpreferred stimuli are probably experienced as neutral or as "sexual but less sexual." For some studies, the researchers do not clearly indicate how they are conceptualizing the nonpreferred stimuli (e.g., Safron et al. 2017). In others, they identify the nonpreferred stimuli as disgusting and/or report that participants found them to be disgust-inducing (e.g., Zhang et al. 2011). This uncertainty about how to interpret nonpreferred stimuli is reflected in the fact that a group of scientists who performed a meta-analysis of these studies originally titled their article "The Neural Correlates of Sexual Arousal and Sexual Disgust," but when the article was published in its final form, it was reconceptualized as "Different Neural Correlates of Sexually Preferred and Sexually Nonpreferred Stimuli" (Long et al. 2019, 2020).

One group of researchers has conducted a number of experiments comparing brain response to erotic pictures and pictures of money (Sescousse, Caldú, et al. 2013; Sescousse, Barbalat, et al. 2013; Sescousse et al. 2015; Gola et al. 2017). These scientists argue that sex and money activate a common reward system but also identify "reward-specific" areas of the brain. They assert that sex is a "primary" reward, whereas money is a "secondary" one, and that primary reward processing likely developed earlier in evolutionary history and may occur earlier in individual development (Sescousse et al. 2010, 13102).

Although, on the one hand, researchers seem intent on performing ever more studies to isolate sex from nonsex, there are a number of what seem like obvious comparisons that have not been directly made in the same study. For one, some have measured brain response to pictures of the subject's romantic partner(s) (which the scientists call the neuroimaging of love or romance) (e.g., Fisher et al. 2005; Bartels and Zeki 2000). However, no studies have directly compared brain activation to erotic stimuli vs. romantic stimuli, although at least one meta-analysis has been performed on this question (Cacioppo et al. 2012), nor have any directly compared brain activation to erotic stimuli vs. food stimuli, although several meta-analyses have been conducted (e.g., Sescousse, Caldú, et al. 2013). In online asexual communities, visitors and new members are often greeted with textual or visual cake, thus humorously comparing the pleasures of eating cake to those of sex.[34] At least from an ace perspective, then, this is a "natural" comparison to make, but not one that researchers have made.

Thus, while I agree with Dussauge that researchers do seek to contain an epistemological anxiety of specificity by performing ever more nuanced comparisons and analyses, I also argue that the frequent selection of "neutral" stimuli as a comparison, plus a failure to perform some comparisons, such as to romance or food, suggests that this field may be haunted by the lingering fear that perhaps sexuality is not so specific after all. If I were required to design a neuroimaging study to isolate brain sexual response, I would probably start by asking subjects to choose their own stimuli, identifying images or videos that evoke, for them, sexual arousal, feelings of romantic or familial love or friendship, hunger or thirst, playfulness or pleasure in physical movement, softness of touch, fear of violation, greed or ownership, and bodily disgust, among others. However, even just beginning to conceptualize such an experiment reveals the limitations of this endeavor: if I conducted this experiment and then subtracted brain activation in response to these other stimuli from that in response to the erotic stimuli, I might identify some areas uniquely activated or deactivated by erotic images—but what meaningful information would this provide? It would not tell us about all of the different brain systems or processes involved in sexual arousal or desire, nor would it necessarily identify the most important ones involved. In addition, as noted above, the subtraction method serves to filter out the multifaceted thoughts and feelings that might be triggered by erotic stimuli in favor of isolating what is unique to sexuality. In sum, my argument here comes down to two somewhat contradictory claims: (1) scientists seem committed to isolating what is specific to brain sexual response but perhaps (unconsciously) avoid robust comparisons due to a fear of undifferentiated desires and pleasures; and (2) subtraction is not a useful or meaningful approach to understanding sexuality and brains.

This review of the why and the how of neuroimaging research on sexual response, arousal, and desire reveals three fundamental (and problematic) assumptions that this research relies on and reproduces: (1) brain sexual response is universal (everyone has the same basic system) and, for many researchers, innate; (2) sex is holographic—scientists can examine one single example or piece of it (brain activation while viewing stock photos of naked bodies or people having sex) and understand it in its entirety ("the brain sexual response system"); and (3) sex is unique—we can subtract away all that is "not-sex" or, more generously, not-just-sex, in order to arrive at "sex" or "just-/only-sex," and the unique part of sex is the part we should be most interested in.

This review does suggest that researchers are able to utilize neuroimaging technology to produce "good" data about certain limited aspects of sexuality: brain activation during sex acts that can be performed during a scan and, especially, while viewing normatively erotic photos or videos in a lab. And, at least among the research subjects studied thus far (which, as a reminder, are disproportionately white, male, heterosexual young adults), brain activation while viewing vss seems similar. If this were all that scientists were claiming to do, the limitations of these studies would perhaps not matter so much. But because so many are focused on "real world" applications—particularly treatments for so-called sexual disorders—the lack of ecological validity becomes glaringly problematic. How a brain responds to passively viewing normatively erotic pictures or videos of unknown (or even composite) people in a lab may have little to do with how a person does or does not experience sexual arousal or desire, or makes decisions about sexual activity, in the context of their actual lives, when contextual factors—including structural, situational, and relationship ones—usually play a significant role.

Background Assumptions Revisited: Models of Sexual Desire, Arousal, and Response, Part II

As discussed in the next section, the incentive motivation model of sexual desire is a fairly nonessentialized one. However, despite the ostensibly widespread adoption of this model and its consistency with the neuroimaging results discussed above, it is clear that a number of scientists in the field retain a "drive" model of sexual desire. Partly this is demonstrated by the controversy over "intrinsic" vs. "responsive" desire, discussed extensively by Alyson Spurgas (2020). To briefly summarize, Helen Singer Kaplan, who largely retained a Freudian understanding of sexual desire as a drive, conceptualized a difference between spontaneous or intrinsic desire and "responsive" desire. According to sexologists, only spontaneous desire was incorporated into the conceptualization of sexuality utilized in the *Diagnostic and Statistical Manual of Mental Disorders*. However, beginning in the 1990s, a number of sex researchers began to maintain that women, in particular, often develop feelings of desire in response to partner initiation of sexual activity (responsive desire). In what they saw as an effort to avoid the pathologizing of "normal" female sexuality, for the *DSM-5*, they created a diagnosis of "female sexual interest/arousal" disorder (to diagnose women only when they report no spontaneous *or* responsive desire), while "hypoactive sexual desire disorder" was retained for men only (if they report no spontaneous desire). However, as Marta Meana (2010) points out, in keeping with an incentive motivation model, all sexual desire is in some way "responsive," developing in response to environmental cues, although some may be more "internal," such as bodily sensations or thoughts, while others may be more "external," such as people, places, videos, and so on. Arguably, in holding on to the distinction between spontaneous and responsive desire, some psychologists and neuroscientists demonstrate their continued (perhaps unconscious) commitment to a drive model of sexuality as opposed to an incentive motivation model. This may also explain the phenomenon noted above, namely, that scientists have not directly performed many of the comparisons immediately suggested by the incentive motivation model, such as comparing sexual motivation to hunger, thirst, or drug-craving.

A minority of researchers in the field assert explicitly that the neuroimaging research on sexual arousal, response, and desire actually provides support for Freud's theories of sexuality. For example, while Lorenzo Moccia et al. (2018, 5) argue that modern neuroscience research has demonstrated that Freud was incorrect in arguing that all motivational behavior can be explained by the original libidic drive, they also claim that the brain's "SEEK-ING/desire system" (identified by modern neuroscience) "displays a series of analogies with the Freudian concept of *libido*." They continue: "The activity of the SEEKING system, in fact, promotes an appetitive predisposition in individuals, a euphoric mental state that is itself gratifying, which is thought to allow individuals to enter into relation with the surroundings in positive affective terms." As another example, Serge Stoléru (2014) states that the modern "neurophenomenological model" of sexual arousal identifies the neural correlates of the sexual drive as described by Freud. Specifically, according to him, the neurophenomenological model has identified the neural correlates of the following aspects of Freud's model: the sexual object, the motor factor, the sexual aim, the source of the sexual drive, and tension (unpleasure) and the elimination of tension (pleasure). The major difference between Freud's understanding of the sexual drive and the neurophenomenological model is that psychoanalysis locates the origin of the former in the peripheral organs, while neurophysiology locates its origin in the brain. Thus, while most scientists in the field have at least ostensibly moved away from a drive model of sexuality, it persists among some in the field. In general, the commitment of neuroimaging research to identifying a single, universal "sexual response system," plus the lingering conceptualization of this system as an innate drive, reflects and reinforces compulsory sexuality and the pathologization of nonsexualities. As discussed in the introduction and chapter 1, a universal drive model of sexual desire promotes the marginalization and stigmatization of asexuality, as, in this view, the absence of sexual desire or attraction appears only as a disorder or symptom of repression. Indeed, as argued above, one of the explicit aims of neuroimaging research on sexuality is to develop "treatments" for hypoactive sexual desire disorder (see chapter 1 for a further discussion).

Thinking Sexual (Non)Attraction Developmentally

Reading the neuroimaging research on sexuality through the analytic of compulsory sexuality demonstrates that this research is still informed by a drive model of sexual desire and a pathologized understanding of nonsexualities. However, "rereading" this research through the analytic of asexual possibilities denaturalizes commonsense understandings of sexual pleasure and desire by suggesting that perhaps they are not nearly as distinct from other pleasures and emotional and motivational states as is commonly thought, and that sexual desire is not the intrinsic motivating drive suggested by the drive model of sexuality. As noted above, many scientists admit that there is no clear distinction between sexual arousal and desire (hence the use of the phrase "sexual response system"). In addition, a number of neuroscientists believe that evolution tied sex (specifically genital stimulation) to a common pleasure/reward system that can be activated by a variety of stimuli.[35] Thus, from the perspective of asexual possibilities, genital stimulation can be conceptualized as simply one among many activities that produce pleasure, primarily distinguished by the connection to reproduction, the specific body parts involved, and (for humans and some other animals) the social meanings attached. If humans experience some types of pleasurable bodily stimulation or contact beyond the genital as sexual but not others, then this difference between "sexual" and "not-sexual" is a constructed difference, not an essential one. Similarly, the neuroimaging data suggest that "sexual motivation" is akin to other motivational states and feelings. Again, from the perspective of asexual possibilities, sexual motivation becomes simply one among a number of motivational states, made "special" primarily by the social meanings attached.

As limited as it is, the neuroimaging research reviewed also seems to provide support for—or at least does not contradict—the incentive motivation model of sexual desire.[36] In a drive model, sexual desire is assumed to be a unique instinct intrinsically produced by all "normal" brains, all of which will naturally develop the capacity to produce and experience it. In contrast, the incentive motivation model is universalizing, but not particularly essentializing, apart from assuming that, as a result of evolution, brains have the capacity to recognize and evaluate stimuli and make decisions about whether (and how) to pursue and/or avoid stimuli. This model allows for the possibility that it is only through learning that people come to categorize certain stimuli as "sexual," and central to it is the idea that how people respond, both consciously and unconsciously, to stimuli categorized as sexual depends a great deal on memories and past experiences, as well as social norms and intersecting systems of oppression. Thus, rereading this neuroimaging research through the analytic of asexual possibilities contributes to a disruption of the sexual/nonsexual binary, thus clearing the way for more flexible, embodied, and socially situated understandings of (non)sexuality. Elsewhere (Gupta 2022) and in the conclusion of this book, I argue that combining the incentive motivation model with a developmental approach can produce an ace-positive understanding of sexual/nonsexual development.

The next chapter of the book, "Pandas, Voles, and Rams: Asexual Phenomena in Nonhuman Animals," examines scientific research on sexual activity among nonhuman animals through the analytics of compulsory sexuality and asexual possibilities. Like the neuroimaging research on sexuality, scientific research on sexual (non)behavior in nonhuman animals reflects and reinforces compulsory sexuality, in this case through its limited scope, focus on pathology, and use of pejorative language (e.g., the term *duds*). However, rereading this research through the analytic of asexual possibilities again contributes to blurring the sexual/nonsexual binary, in this case through denaturalizing the category "sexual activity/behavior."

CHAPTER 3

Pandas, Voles, and Rams

Asexual Phenomena in Nonhuman Animals

"PANDAS ARE CUDDLY, BUT NOT TO EACH OTHER. THEY MUSTER ABOUT as much enthusiasm for sex as a human does for a root canal," reads a *New York Times* article titled "Lousy Libidos: Why Do Pandas Have So Little Sex?" (E. Wong 2016). "There is perhaps no mammal that is less often in the mood for sex than the female giant panda," reads a *Scientific American* article (Jabr 2012). According to these and others, scientists struggle to breed giant pandas in captivity, which is variously attributed to: that female pandas only enter estrous once a year for 24–48 hours, that male pandas have a small penis relative to their body size, and/or the speculation that pandas living in the "luxury" of zoo habitats have become too "lazy" to mate (Manevski 2021; Castro 2016). Some commentators claim that giant panda sex drives are so low, it is a miracle that the species has avoided complete extinction (Buchen 2008). The media seem to enjoy covering the trials of giant panda sex in captivity—especially stories about Chinese scientists dosing male pandas with Viagra and providing them with "panda porn" in an effort to facilitate sex (Getzlaff 2000). In 2017, Pornhub, one of the largest internet porn sites, asked people to dress up as pandas and film themselves having sex in order to generate more porn for captive pandas and to raise money for panda conservation (Chatel 2017).[1] Human visitors have flocked to zoos during "breeding season" in a bid to view panda sex in person (Uddin 2013), although some zoos now close viewing areas and turn off live-streaming panda webcams during this time (Gannon 2015).

Perhaps not surprisingly, the media stories about giant panda sexual inadequacies attracted the attention of some ace people. A 2016 thread on AVEN begins with a post lightheartedly suggesting that pandas would make a good (gray-)asexual symbol because of their low sex drives (AVEN, n.d.-c).[2] In other online discussions about animal symbols of asexuality, pandas are usually mentioned as contenders (along with frogs, jellyfish, and axolotls), although dragons seem to be the most popular animal symbol of asexuality at present (see, e.g., "R/Asexuality," n.d.; Casey 2018; ladypoetess, 2011; AVEN, n.d.-b). At the time this was written, ridiculously adorable asexual panda-pride products could be purchased from Amazon and Etsy, among other sites. As a possible tongue-in-cheek animal symbol of asexuality, pandas have a lot to recommend them, as they are well known for their sexual difficulties but also adored for their lovable squishiness.

However, as some media outlets and ace people acknowledge, the truth about giant panda sexuality is both more complicated and much bleaker than is suggested by stories about panda porn and sexually incompetent males. According to cultural critics, giant pandas began generating interest world-wide in the late nineteenth century but only became a national icon in China in the 1950s, after they were adopted as a symbol by the Maoist government, a move made in part because pandas had never been used in that way by China's imperial dynasties (Songster 2018). However, due to habitat destruction, panda populations were declining. The giant panda was adopted by the World Wildlife Fund for its logo in 1961, becoming an international symbol of species endangerment and conservation (Nicholls 2010). The United States formally listed the giant panda as an endangered species in 1984, noting a total population of a thousand (Koerth 2017).

To its credit, beginning in the 1960s, the Chinese government moved to protect panda habitat but also began a concerted effort to breed the animals in captivity (Songster 2018). It also began to send pandas to zoos in selected countries (originally as gifts, later as loans, often in exchange for significant compensation)—a practice labeled "panda diplomacy" by observers (Barua 2020; Collard 2013). Zoos all over the world took on the substantial costs of hosting pandas, because they were expected to be income generators—predicted to attract corporate sponsors, entice millions of visitors, and drive the purchase of millions of dollars' worth of panda-related merchandise (Mott 2006; Barua 2020). Over time, a worldwide network emerged of scientists, conservationists, and donors dedicated to panda breeding in captivity. For many years, these efforts were unsuccessful, but not, as some media articles claimed, solely as a result of inherent panda deficiencies—apparently pandas living in the "wild" manage to breed just fine (F. Wei, Hu, et al. 2015; Nicholls 2010; Koerth 2017). Instead, the created habitats of zoos restricted important aspects of panda sexuality—including opportunities for young pandas to observe adult ones having sex, for pandas to mark their territory with their own scent and smell the scent marks of other pandas, and, perhaps most importantly, for pandas of both sexes to choose from among a group of possible partners (Koerth 2017; Uddin 2013). Lisa Uddin (2013) describes the staged nature of panda mating at the National Zoological Park in Washington, DC, in the 1970s and 1980s, noting that staff attempted to present a heterosexual reproductive romance between an adult female and an adult male panda, to resounding failure. She describes this reproductive failure as productive of queer possibilities, particularly when a second male was brought into the drama, an effort that had the same result. Some might be tempted to read this failure as productive of *asexual* possibilities in particular, but if so, these are possibilities produced by a variety of traumas, including human-caused habitat destruction and the carceral logics of zoos.[3] Zoos' giant panda breeding programs have also been criticized for their methods of artificial insemination—a process that generally involves restraining the male and female pandas and using electroshock to cause the male to ejaculate (Koerth 2017).[4] Finally, a number of commentators have argued that the massive amount of money spent on captive panda breeding programs would be better spent on other conservation efforts (e.g., Dell'Amore 2013).

Since the 1990s, zoos and conservation programs have implemented new practices to promote panda breeding in captivity, based on an increased understanding of panda biology and behavior (Martin-Wintle et al. 2015; F. Wei, Swaisgood, et al. 2015). Providing pandas with a choice of mate, in particular, appears to significantly increase at least the rate of copulation, from as low as 0 percent when neither panda is a "preferred" partner to near 80 percent when both are "preferred" (Martin-Wintle et al. 2015).[5] As a result of various "improvements," captive breeding rates have increased significantly since the 1990s, although the media often continues to represent panda mating in captivity as difficult and rare (e.g., Manevski 2021; E. Wong 2016). From the 1960s until the mid-1990s, only around 115 cubs were born in captivity, and only 16 of these survived to reproductive age (Nicholls 2010, 204). In 2017 alone, 63 giant pandas were born in captivity (Zhiling 2017). Due to the increase of pandas living in the wild (up to around 1,800 in 2021) as well as in captivity (over 630 in 2021), in 2016, the International Union for Conservation of Nature changed the status of the giant panda from endangered to vulnerable, a change that was endorsed by China in 2021 (Pruitt-Young 2021; Obermann 2021). Now, commentators are concerned about how many giant pandas are living their entire lives in captivity and whether many will ever be successfully released into the "wild" (Koerth 2017).

This story about giant panda (a)sexuality previews many of the themes of this chapter. Perhaps most importantly, it demonstrates the continuing fascination that humans have with the (a)sexuality of nonhuman animals, the value judgments they make about it, and the ways in which nonhuman animal (a)sexuality and reproduction is often materially entangled with human ecological destruction and economic and political desires. On the one hand, some ace-identified people are able to find validation in and identify with sexually uninterested pandas. On the other, at least some of that disinterest is the result of humans disrupting their physical and social environments. In addition, in the media derision of and humor at panda mating failures, we can see the operation of both compulsory heterosexuality and compulsory sexuality. The failure to have sex and reproduce is at best an occasion for humor and at worst a reason for writing off a whole species. Significant time, energy, and money have been spent by conservationists, zoos, and the Chinese government to overcome this nonsexuality and nonreproductivity (sometimes through force, as in the case of artificial insemination). I find myself particularly troubled by the racial, ethnic, and national resonances of these discussions: While I did not find any evidence in media articles that people were connecting panda sexuality to a specifically Asian or Chinese sexuality, it is also true that representations of the panda as sexually incompetent or disinterested echo closely Western representations of Asian and Asian American men as effeminate and asexual (Le Espiritu 2004; Eng 2001; C. Han 2006; Shek 2007; Nguyen 2014). Even if stereotypes about Asian male sexuality are not motivating a focus on the sexual incompetence of pandas, the fact that the giant panda has been adopted as one of China's national symbols means that discussions of its sexuality are inextricably linked to those about the sexual character of Chinese nationals.

This chapter takes up some of these themes in different ways, using the analytics of compulsory sexuality and asexual possibilities to read scientific research on (a)sexuality in nonhuman animals, as well as its social impact.[6] Such research has long played a role in public debates about sexuality, with some scientists and activists countering claims that homosexuality is "unnatural" by asserting that evidence for it in nonhuman animals proves that it is "natural" in humans. As mentioned above, evidence for nonsexuality in nonhuman animals has already been incorporated into discussions about human asexuality. Reading scientific research on nonsexuality in nonhuman animals through the analytic of compulsory sexuality reveals that it reflects and reinforces compulsory sexuality through its limited scope, its use of pejorative language—such as the term *duds*—and its focus on medical intervention, which reflects what disability studies scholar Rosemarie Garland-Thomson (2012) refers to as "eugenic logics."

However, I also argue that although our identification of asexual-like phenomena in nonhuman animals can never be impartial or disinterested, scientific evidence of such can render strange our own problematization of asexuality among humans. I then return to the question of whether ace and queer activists should celebrate cases of so-called aceness or queerness in nonhuman animals when it seems to be the result of environmental destruction, using Jasbir Puar's concept of debility to clarify the stakes involved. I argue that, rather than using evidence of non- or queer sexuality in nonhuman animals to validate our political projects, we can "reread" scientific findings about sexuality in nonhuman animals through the analytic of asexual possibilities to further disrupt the sexual/nonsexual binary, in this case, by unsettling the category of "sexual activity or behavior."

I begin by reviewing foundational queer feminist critiques of scientific research on sexuality in nonhuman animals. I then draw from Angela Willey's work on compulsory monogamy to offer an example of the influence of compulsory sexuality on research on animal sexuality. In her analysis of research on vole monogamy, Willey suggests that compulsory sexuality led scientists to interpret social behavior among voles as necessarily sexual. I then offer an original analysis of other scientific studies on "asexual phenomena" in nonhuman animals.

Before proceeding, it is important to note that the distinction between "human" and "nonhuman" has always been a political one. As scholars have attested, people of color, people with disabilities, and other minoritized groups have long been denied inclusion in the category of "human" by dominant groups, ideologies, and institutions (McKittrick 2015; Parker 2018; M. Chen 2012). European scientists in the eighteenth and nineteenth centuries claimed that Africans and other people of color were a separate (and inferior) species (Jackson and Weidman 2004; Dennis 1995; Schiebinger 1993). In the US, mainstream discourse continues to depict Black, Indigenous, and other people of color, as well as people with disabilities, as "less human" and "closer to animals" than white, able-bodied people (M. Chen 2012; Weaver 2021). Commentators have also contended that the human/nonhuman distinction reinforces speciesism (Horta and Albersmeier 2020; Wyckoff 2014; Westerlaken 2020). In this book, I use the terms *human* and *nonhuman animals* to indicate both connections and disconnections between them. I do not intend to imply a hierarchical relationship between them; however, I recognize that pairing these terms may suggest a binary opposition, simultaneously erasing commonalities and differences within the categories of "human animals" and "nonhuman animals."[7]

Foundational Queer and Feminist Critiques of Animal Sexuality Research

Feminist observers have long argued that what scientists are capable of noticing about the behavior of male and female animals is shaped by social and cultural expectations about gender, sexuality, and other systems of inequality, and that it is important to analyze research on nonhuman animals from an intersectional perspective (Taylor 2024). Regarding sexuality specifically, in the 1970s, feminists and their allies inside and outside of the field of animal behavior criticized scientists for their inability to see the sexual behavior of female nonhuman animals as anything other than passive. Beginning in the 1970s, careful observation combined with different expectations about masculine and feminine sexuality allowed animal researchers to observe female animals pursuing and initiating sexual activity (Fausto-Sterling 2000; Haraway 1990).

Feminist and queer studies scholars have been especially concerned about scientific studies involving nonhuman animal models to understand human sexual orientation, particularly homosexuality, along with the political ends to which they have been put (e.g., Adriaens and Block 2022; Terry 2000). Although the concerns are varied, here I group them into three broad categories. The first is largely directed at how scientific evidence about nonhuman animal sexuality has been used. Understandably, because homosexuality has often been dismissed as "unnatural," scientists and LGBTQ+ activists have utilized evidence of "homosexual" behavior among nonhuman animals to support their belief that human homosexuality is "natural" and, by implication, worthy of tolerance. Feminist and queer critics point out that doing so is based on a faulty equation of naturalness with goodness (e.g., Walters 2014). In addition, commentators point out that scientists and activists are often reading their own political agendas into the behavior of nonhuman animals, even when the behavior may not fully align with a particular narrative. For example, Noel Sturgeon (2010) contends that both US conservatives and liberals have employed penguin sexuality to support their own agendas—the former used evidence for the supposed monogamy of emperor penguins to support their own "family values" agenda, while the latter used evidence of the bonding of same-sex penguin pairs to support their push for the legalization of same-sex marriage in the 2000s and early 2010s. As Sturgeon suggests, neither narrative is fully supported by penguin behavior itself—for example, the relatively equal division of reproductive labor among emperor penguins does not seem to be in line with the patriarchal family structures supported by "family values" advocates. At the same time, when pair-bonds between same-sex penguins break down, this can interrupt the positive messaging of same-sex marriage advocates. For instance, Roy and Silo—who formed one of the most famous same-sex penguin pairs—eventually separated, and Silo entered a relationship with a female penguin. In fact, says Sturgeon, the evidence suggests that penguins may not be uniformly monogamous after all, whether in different-sex or same-sex relationships: "Arguments from the natural about sexuality, of whatever kind, especially when one uses penguins as one's touchstone, turn out to be pretty slippery" (113). As we saw at the beginning of this chapter, while pandas may be an initially appealing symbol for ace-identified people, the reality of their sexual practices in the wild as well as the ways in which their sexuality is curtailed in captivity make such an identification problematic. In addition, as white supremacist societies associate Black, Indigenous, and other people of color with nonhuman animals as a form of symbolic violence (M. Chen 2012; Weaver 2021), a political argument based on drawing a connection between ace people and nonhuman animals may not offer the same benefits to ace people of color as to white aces.

The second and third sets of critiques focus less on the political use of scientific evidence and more on the research itself. The former set is that many studies (and their translations in popular media) confuse same-sex sexual activity with gay or lesbian orientation or identity. Queer and feminist scholars differentiate between sexual desire (what people want—or don't—sexually), sexual activity or behavior (what people do or don't do sexually), sexual identity (how people think and talk about their own a/sexuality), and sexual orientation, a concept not endorsed by all feminist or queer academics but which Ed Stein (1999, 45) defines as a "disposition" based on a person's "sexual desires and fantasies and the sexual behaviors he or she [*sic*] is disposed to engage in under ideal conditions." As feminist and queer scholars maintain, one's (a)sexual identity does not necessarily depend on either desires or behaviors. Yet, in many scientific studies on animal sexuality, researchers observe same-sex sexual behavior among nonhuman animals and then equate this to human gay or lesbian identity or orientation. In some cases, this is same-sex behavior that is produced in the lab as a result of brain surgery, exposure to hormones, or atypical rearing environments.[8] In addition, in laboratory settings it is rare for nonhuman animals to be given a choice of sexual partners. As a result, it should not be argued that the same-sex sexual behavior observed among them, particularly in the lab, is necessarily similar to human gay or lesbian sexual orientation or identity (Adriaens and Block 2022; Alaimo 2016; Vasey 2002; Terry 2000). In fact, as Jennifer Terry (2000) notes, in some of the early research on male rat "homosexuality," the subjects were not even necessarily engaging in sexual behavior with other males but were instead displaying what is called "lordosis behavior," which is a "female-typical" sexual behavior in rats. In this case, male rats displaying this behavior were identified as "homosexual," regardless of the sex of their sexual partner(s) (see also Pettit 2012; Jordan-Young 2010, especially chapter 7).

Some animal researchers have responded to these criticisms. For example, Paul L. Vasey (2002) agrees that evidence of same-sex sexual behavior in nonhuman animals cannot be taken as evidence of homosexual identity (as they are unlikely to develop sexual identities). He also agrees that such evidence, when it is produced by brain lesions, hormone exposure, or atypical rearing environments, or when it is only observed in the absence of mate choice, should not be taken as evidence of homosexual orientation. Vasey develops more stringent criteria to evaluate evidence of homosexual orientation in nonhuman animals: Specifically, the animal must be reared in a species-typical environment and must not be subject to surgical or hormonal interventions. It must also be given the opportunity to interact sexually with both male and female conspecifics in a situation in which it has control over its own sexual activity. If these criteria are met, and the animal is observed to express a preference for interacting sexually (and not just socially) with one sex but not the other, only then can it be described as having a homosexual orientation.[9] Using this stricter definition, Vasey finds evidence of "homosexual" orientation among female pukekos (a type of bird), cows, domestic rams, female Uganda kobs (a type of antelope), and female Japanese macaques (a type of monkey). However, while Vasey's stricter definition of homosexual orientation among nonhuman animals responds to feminist and queer concerns that emphasize the disarticulation of desire, behavior, identity, and orientation, it does not answer the question of why it matters whether homosexual orientation exists among nonhuman animals.

Third, a number of feminist and queer critics have maintained that animal models should not be employed by scientists to understand human sexuality, because, in this view, the influence of social, cultural, economic, and political structures may make human sexuality qualitatively different from that of nonhuman animals (discussed in Adriaens and Block 2022).[10] Consider the example of using animal models to develop drug "treatments" for human sexual "disorders." In their article "What Can Animal Models Tell Us About Human Sexual Response?," James G. Pfaus et al. (2003, 6) argue that the validity of a comparison between animal sexuality and human sexuality "can only be determined in situations that test whether a treatment that modifies behavior in the animal does so in humans." They point out that many of the drugs that enhance penile erection in male rats (e.g., sildenafil) also do so in male humans. Thus, they contend that it is possible to learn something meaningful about human male sexual arousal from experiments with male rats. While feminist and queer analysts would not necessarily dispute the claim that similar biological processes underlie penile arousal in rats and humans, they would suggest that, in humans, (dys)function is often not simply a biological event but a political, social, and interpersonal phenomenon. While prevalence estimates vary widely, medical researchers argue that a significant percentage of cases of erectile dysfunction are primarily "psychogenic" in origin, particularly among younger men (Ciaccio and Di Giacomo 2022). At the very least for these cases, the political, social, and interpersonal aspects of penile (dys)function must be studied and taken into account in any effort to address the issue at either the individual, interpersonal, or societal level (if, indeed, we want to address it at all). In other words, per these critiques, studying male rats may not give us particularly meaningful knowledge about many cases of human erectile (dys)function.

Compulsory Sexuality in Scientific Research on Vole "Monogamy"

Angela Willey's (2016) analysis of research on vole "monogamy" provides evidence of the ways in which compulsory sexuality can influence scientific studies of the behavior of nonhuman animals (see also Willey and Giordano 2011; Gupta 2012b). In her work, Willey analyzes research carried out at Dr. Larry Young's neuroscience lab at the Yerkes National Primate Research Center in Atlanta, Georgia, on "social monogamy" in prairie voles. As Willey documents, this research, which gained national attention in the 2000s, sought to identify the genetic and hormonal causes of "social monogamy" in a particular species of prairie voles. Willey notes that this species was chosen to serve as an animal model for humans because the researchers believed that both species are "socially monogamous" (by which they signify pair-bonding and coparenting, not sexual fidelity). As Willey points out, this essentializes monogamy as the natural form of human relationality. In addition, she identifies the problematic slippages that took place both in the lab and in media reports of its research: In both, social monogamy is equated with the ability to form "healthy" relationships, and thus social monogamy became equated with healthy sociality while nonmonogamy was associated with asociality or social dysfunction. The researchers sought to identify genetic and hormonal interventions that could transform nonmonogamous voles into monogamous ones, in the hopes that these "treatments" could eventually be used in humans. In particular, because the researchers believed that autism is characterized by social dysfunction, they argued that "treatments" used to produce social monogamy in prairie voles could be employed to "treat" autism in humans. As Willey points out, these assumptions reflect and reinforce compulsory monogamy, heterosexism, racism, and ableism.

In addition, Willey identifies the influence of compulsory sexuality on this research. She notes that, in the lab, a male and female vole are left together for eighteen hours unobserved, during which time they are presumed to mate and possibly pair-bond. After that, the male is subject to a partner-preference test: He is placed in an enclosure with the familiar female and an unfamiliar female, each tethered to a side of the cage. The researchers then measure how much time the male spends with each, which they call "huddling." If he spends enough time with the familiar female, they consider him to be "socially monogamous." However, when Willey observed the experiment, she noted that the animals did not seem to be "huddling" or having sex; in fact, the male voles seemed to be trying to chew through the tethers binding the females in place, while the latter seemed to be trying to help the males remove the shunts (small tubes) that had been surgically inserted into their heads. According to Willey, the behavior could be described as a form of sociality, solidarity, cooperation, resistance, or even bonding, but not necessarily as romantic or sexual.

Based on the fact the initial bonding time is unobserved and on her own observations of the partner-preference test, Willey points out that the researchers don't actually know that sex is producing the pair-bond or that the bond formed is sexual or romantic in nature. Willey writes, "A closer look at what is being measured suggests that sexuality is an interpretive frame imposed on pair-bonding behaviors" (2016, 60). She also points out that the lab was not testing bonding between male-male or female-female pairs, or bonding between more than two voles. Willey reasons that the researchers see the ultimate basis or exemplar of human sociality to be a monogamous, heterosexual, and reproductive pair-bond. Thus, they set up experiments that sought to artificially induce and then test for this specific type of pair-bonding among voles, rather than sociality, bonding, or relationality more broadly. And then, regardless of what actually occurred between the voles in the lab, their behavior was interpreted as sexual pair-bonding. As Willey writes, "The scientific naturalization of monogamy in this laboratory reinforces not only the 'twoness' requirement for relationships but also the idea that the human is fundamentally sexual. . . . The special status of coupling [in this research] depends on distinguishing it not only from casual or uncoupled sex but also from friendship, comradeship, and situational solidarity. Breaking down barriers between these naturalized categories has the potential to radically reshape how we understand the importance of sex to human nature. It calls us to rethink the pervasive cultural privileging of sexual relationships over other types of connections" (72).

Analyzing Scientific Research on "Asexual Phenomena" in Nonhuman Animals

Drawing loosely on Vasey's (2002) discussion of homosexual orientation in nonhuman animals discussed above, I define evidence of "asexual phenomena" in nonhuman animals as evidence of those who do not engage in sexual activity when given the ability to control their own sexual behavior. I call this not "asexuality" but "asexual phenomena" to emphasize that what we observe for nonhuman animals is not the same as asexual identity in humans. On the surface, this behavioral definition does not align with definitions of human asexuality, which focus on a lack of sexual attraction to others and consider it irrelevant whether ace people engage in sexual behavior or not. Arguably, unlike the neuroimaging research examined in the previous chapter, which focused on accessing interior psychological states, the animal behavior research is primarily focused on "exteriority" or observable behavior. However, some animal behaviorists maintain that under controlled conditions, internal motivation can be inferred based on observed behavior.[11] According to this then, observing an absence of sexual behavior by nonhuman animals when they are in control of their choices could lead to the inference that they lack sexual motivation; by this reasoning, a behavioral definition of asexual phenomena in nonhuman animals may be closer to the human definition of asexuality than it initially appears. Here I restrict myself to examining scientific research on asexual phenomena in mammals, which has provided evidence for asexual phenomena in rodents, rabbits, ungulates, and nonhuman primates.[12]

Male Rats and Mice: Since the 1960s, animal researchers have been aware that some male rats do not engage in sexual activity when provided with the opportunity to do so. These are called "noncopulators" (Whalen et al. 1961). A research team at the Universidad Nacional Autónoma de México led by Wendy Portillo and Raúl G. Paredes has performed a number of experiments over the past twenty years comparing male rats they call sexually active, "sexually sluggish" (those who engage in sexual activity but are slow to ejaculate), and "noncopulating" (those who do not engage in sexual activity). They report that approximately 3 percent of male rats do not engage in sexual behavior (Portillo et al. 2003), although other researchers have found rates of noncopulation to be as high as 20 to 40 percent (Canseco-Alba and Rodríguez-Manzo 2019). Based on previous research along with their own studies, Portillo and Paredes argue that noncopulating male rats have circulating levels of testosterone that fall within the species-typical range and that administering testosterone to them does not increase their sexual behavior (summarized in Ventura-Aquino and Paredes 2017). However, they found differences between noncopulating and copulating male rats in the number of specific receptor cells found in areas of the brain related to sexual activity (Portillo et al. 2006). They also found that noncopulating males had decreased aromatase activity (conversion of testosterone to estradiol) in the same areas (Portillo et al. 2007). Additionally, they found that they could induce sexual behavior in noncopulating male rats by implanting testosterone or estradiol directly into an area of the brain related to sexual activity (the MPOA, or medial preoptic area) (Antonio-Cabrera and Paredes 2014). They were also able to induce sexual activity through "kindling," or repeated electrical stimulation of the MPOA (Portillo et al. 2003). Other researchers have been able to induce sexual activity in some noncopulating male rats by administering naloxone (an opioid antagonist), anandamide (a neurotransmitter), or *Phlegmariurus saururus* extract (from a plant commonly used as an aphrodisiac) (Canseco-Alba and Rodríguez-Manzo 2019; Birri et al. 2017).

Portillo et al. (2013) repeated some of these experiments with male mice. They reported that, as in noncopulating male rats, noncopulating male mice appear to have circulating levels of gonadal hormones that fall within species-typical range, and administering estradiol did not induce sexual activity. They also concluded that the lack of sexual motivation observed in noncopulating male mice may be due to "deficiencies" in odor processing.

As discussed below, there is little research on "asexual phenomenon" in female nonhuman animals, but some researchers have studied intraspecies variation in sexual behavior among female rats in an effort to use them as a model to study hypoactive sexual desire disorder (see chapter 1). For instance, Eelke M. S. Snoeren et al. (2011) found that, based on paced mating tests, female rats could be divided into three groups: those that mostly avoided the male, those that mostly approached the male, and a large middle group, and that these preferences seemed to be stable over time. They suggested that the group that mostly avoided the male could be a model for HSDD in women. However, according to Vasey's strict definition, from an animal behaviorist perspective, they would not meet a conservative definition of "asexual phenomenon" in nonhuman animals, as all of them were treated with estradiol to induce sexual behavior.

Male Guinea Pigs: Researchers have found that male guinea pigs also vary in their levels of sexual interest. Cheryl F. Harding and Harvey H. Feder (1976) classified male guinea pigs as "high drive," "medium drive," or "low drive" animals. In an experiment, they found that high-drive and low-drive guinea pigs did not differ in their resting levels of testosterone; however, after exposure to an estrous female, high-drive animals showed slight increases in testosterone levels, while low-drive ones showed slight decreases. As a result, they concluded that "the manner in which individual animals perceive the testing situation may affect their endocrine function, both gonadal and extragonadal, and possibly their neural activity as well" (1205). The low-drive animals were not tested with male guinea pigs, so it is possible that they may have been sexually interested in male conspecifics. I do not know if studies have been conducted to test if, when given access to both male and female conspecifics, there are still male guinea pigs who express low levels of sexual interest.

Male Gerbils: Male gerbils also vary in their sexual and reproductive behavior. Mertice M. Clark et al. (1992) classified males as "studs" or "duds" based on their sexual and reproductive performance; in a 2000 article, Clark and Bennett G. Galef used the label "asexual" to describe male gerbils who express low levels of sexual and reproductive activity. Unlike in male rats and male guinea pigs, Clark et al. (1992) found that differences in sexual and reproductive behavior are correlated with circulating levels of androgens. Clark and Galef (2000) contend that the uterine position of male gerbils (whether they are located next to male or female siblings within the uterus) is correlated with adult levels of circulating testosterone, adult copulatory and reproductive behavior, and adult parenting behavior: As adults, "2F" males (males located between two female siblings in utero) on average express lower levels of circulating testosterone, lower levels of sexual interest and behavior, and higher levels of parenting behavior. They proposed an evolutionary explanation for "asexuality" in gerbils similar to a popular one suggested for "homosexuality" in nonhuman animals: that the latter pass on their genetic material through "parenting" the offspring of their relatives, thus ensuring those offspring are more likely to survive to reproductive age (for a discussion and critique of these theories, see Monk et al. 2019). Applying this theory to asexual gerbils, Clark and Galef write, "For asexual males to enjoy reproductive success comparable to that of sexually active competitors, males' probability of direct reproduction would have to be relatively low and their probability of increasing their inclusive fitness by helping to rear kin would have to be relatively high" (804). Again, it is not clear whether these gerbils were tested for their levels of sexual interest with both female and male conspecifics.

Male Rabbits: Male rabbits demonstrate individual variations in levels of sexual activity. Anders Ågmo (1976) classified male rabbits as expressing either high levels or low levels of sexual activity. He then castrated the animals and tested the responsiveness of these two groups to testosterone propionate (TP). Unlike in male rats and male guinea pigs, he found that, for most measures of sexual activity, administering testosterone produced similar levels of sexual behavior in the two groups. However, the difference in mount frequency remained, even with administration of high levels of testosterone. In addition, the rank order of the animals' sexual activity did not vary with the administration of testosterone. According to Ågmo, his results indicate that a low level of sexual interest in male rabbits is correlated with circulating androgen levels; however, he claims that blood testosterone concentration alone cannot explain the individual variation between male rabbits.

Male Sheep (Rams): Compared to other animals, there has been more scientific interest in individual differences in the sexual and reproductive behavior of domestic farm animals, as these have important economic consequences. As is the case with giant pandas, when they fail to copulate and reproduce, profits suffer. In a review of male animal behavior, Larry S. Katz (2008) described the significant proportion of bulls, rams, and male goats that express low levels of sexual behavior as displaying "sub-standard sexual performance" (he also employed the term *duds*). In addition, he said, attempts to reliably correlate substandard sexual performance with sex steroid levels had not been successful. Here I focus on the case of rams, as the etiology of sexual variation among them has been the most extensively investigated.

Researchers have recognized significant variation in the sexual interests of rams. In a number of studies, they have classified rams as female-oriented (FOR), male-oriented (MOR), or asexual (NOR), the last of which Charles E. Roselli et al. (2002, 264) used if "they demonstrated consistently low levels of courtship with no clear evidence of a preference for either a male or female stimulus animal and no mounting or ejaculatory behavior." The rate of asexuality among ram populations varied from 12.5 percent to 18.5 percent (Roselli and Stormshak 2009). After a number of experiments, Roselli et al. (2002) concluded that differences in circulating androgen levels did not explain the expression of "low libido" in rams. The researchers hypothesized that "neural substrates" mediate the differences between FORS, MORS, and NORS and suggested the following as possible neural substrates that may differ between different groups of rams: aromatase activity, estrogen receptors, behaviorally induced neuronal fos responses, and neuronal cell sizes (see also Roselli and Stormshak 2009).[13]

Male Rhesus Monkeys: Male rhesus monkeys display significant individual variation in their sexual behavior. Phoenix and Chambers (1988) conducted research with five males who were selected for their low levels of "sexual performance"; they had ejaculated in less than 50 percent of previous tests but displayed no observable behavioral pathology. All but one were wild-born. The researchers found that their circulating testosterone levels were within the species-typical range and that treatment with exogenous TP did not change sexual nonresponsiveness (and, in fact, reduced some measures of sexual behavior). However, the males were tested for their sexual interest only with females, not with other males.

Compulsory Sexuality in Scientific Research on
Asexual Phenomena in Nonhuman Animals

As in the case of other scientific research on sexuality in nonhuman animals, studies on asexual phenomena in nonhuman animals reflect a number of social and cultural assumptions about sexuality. For one thing, in the use of terms like *dud* and *substandard sexual performance*, researchers reflect their own valuation of sexual virility in male animals and their associated devaluation of those who express low levels of sexual behavior. In some cases, as in that of giant pandas and domestic farm animals, this greater valuation of highly sexually active males is clearly motivated in part by economic considerations. However, it also likely reflects the equation in mainstream

US society of sexual virility with hegemonic masculinity (itself valued) and compulsory sexuality. Other terms already in circulation, like *noncopulating*, seem less pejorative and better choices moving forward. In at least two cases, researchers have used the term *asexual* to describe rams and gerbils who do not engage in sexual activity. In both, it was employed as early as the year 2000, before its use to describe a human sexual identity or orientation was well established or well known.[14] Using the term for nonhuman animals has benefits and limitations: on the one hand, it can emphasize the ubiquity— and possibly, the nonremarkability—of asexual phenomena in the world (discussed in more detail below), but on the other, it can obscure very real differences between observed behavior (or lack of behavior) in nonhuman animals and sexuality in humans, as discussed above.

One of the most striking findings was the fact that almost all of the research I could find dealt with male animals. There are a number of possible explanations for this. It is, of course, possible that asexual phenomena occur rarely in female nonhuman animals. However, it is also possible they have been observed by researchers but not discussed in the scientific literature because they have not seemed particularly interesting or in need of explanation. Because of social and cultural assumptions about male sexual virility, a lack of sexual interest or behavior in male animals has captured the attention of some researchers and has seemed to require explanation. However, because female sexual disinterest or passivity is often considered "normal" (and does not necessarily prevent reproduction in contexts such as the breeding of domestic livestock, in which male sexual "underperformance" emerges as problematic), asexual phenomena in female animals may not have appeared to require explanation. Perhaps, somewhat ironically, as researchers shift their attention to the expression of active sexuality in female animals, they will come to see asexual phenomena among them as interesting and in need of explanation. As noted above, there have been some studies on female rats that have attempted to identify a subpopulation that could serve as an animal model for hypoactive sexual desire disorder (HSDD) in women. Unfortunately, this research understands sexual disinterest in women through the analytic of pathology and focuses only on its occurrence in female rats for the purpose of developing drug treatments for HSDD (discussed further below).

Another striking finding from this review was another absence: in addition to the lack of research on female animals, there has been a similar lack in regards to asexual phenomena in nonhuman animals in general. This reflects, I believe, a lack of interest in individual variation within populations when it comes to nonhuman animals. Researchers often focus on average levels of sexual interest and activity across a species, or across the males or the females (identifying "male-typical" and "female-typical sexual behavior"), which may obscure asexual phenomena among nonhuman animals. When researchers have focused on individual variation in sexual behavior in nonhuman animals, they have found significant variation, for example, in male rhesus monkeys (Michael and Saayman 1967), female Japanese macaques (Leca et al. 2014), male goats (Bedos et al. 2016), pigs (Hintze et al. 2013), and male camels (Deen 2008), among others. A greater focus on such variations within populations would lead to a more complex picture of nonhuman animal sexuality.

A final striking finding from this review is the way in which medical and "cure" logics are deeply imbedded in scientific studies on (a)sexuality in nonhuman animals. As we have seen, scientists often test pharmaceutical, surgical, or other interventions to see if they change an aspect of the animal's sexual behavior, in the expectation or hope that these will eventually be employed to alter human sexual desire or behavior. This was the case for interventions for erectile dysfunction and research on social monogamy in voles, in which the hope was to develop interventions that could alter social functioning in humans.

On the one hand, the scientific research on asexual phenomena in nonhuman animals could be utilized pragmatically to argue against the use of testosterone to "treat" people who are disinterested in sex. Medical researchers have considered using testosterone to "treat" female sexual desire disorder (Tiefer 2006; Cacchioni 2015a) or "restore" sexual interest in aging men (Marshall and Katz 2013). It seems possible that at least some people who are still in the process of coming into an ace identity might be encouraged or even pressured by partners, family, friends, medical providers, and/or societal messages about sexuality to take testosterone in an effort to increase their interest in sex.[15] Given that the preponderance of the studies reviewed suggests that administering testosterone to "noncopulating" male animals does not alter sexual nonresponsiveness, perhaps this research on nonhuman animals could be drawn upon by ace communities or individuals to protest the use of testosterone to "treat" men with stable, lifelong asexual orientations, on pragmatic rather than moral or ethical grounds.

However, scientists have identified the following methods to induce sexual activity in at least some noncopulating male rats: electrical stimulation of the brain (Portillo et al. 2003); implanting testosterone or estradiol directly into a specific brain area; and administering naloxone, anandamide, or *Phlegmariurus saururus* extract (Canseco-Alba and Rodríguez-Manzo 2019; Birri et al. 2017; Antonio-Cabrera and Paredes 2014). These findings suggest the limitations of making liberatory claims (even pragmatic ones) on the basis of scientific evidence. In reality, all studies on sexuality come with this risk: Some scientists and activists have drawn on research about the "biological basis" (whether genetic or hormonal) of sexual orientation (including evidence for the existence of homosexuality in nonhuman animals) to argue for the innateness, immutability, "naturalness," and therefore benign character of minority sexual orientations. However, this evidence could also always be employed in efforts to develop interventions to alter sexual orientation (such as conversion therapy) and/or prevent people with minority sexual or gender orientations from being born (e.g., through selective abortion) (for a general discussion, see Halley 1994). If noncopulating male rats are seen as analogous, in some way, to ace humans, does that mean that society will consider drug interventions to increase sexual attraction appropriate for ace humans?[16]

In "The Case for Conserving Disability," Rosemarie Garland-Thomson (2012) defines eugenic logic as that which tells us that our world would be a better place if disability could be eliminated. She writes that "eugenic logic is a utopian effort to improve the social order, a practical health program, or a social justice initiative that is simply common sense to most people and is supported by the logic of modernity itself" (340). Much research on animal sexuality and sociality is steeped in this logic—whether it is erectile dysfunction, hypoactive sexual desire disorder, or autism, it seeks to use animal experimentation to develop "treatments" to eliminate various forms of disability from the world. In addition, the goal of reducing or eliminating disability in humans seems to justify killing (researchers call this "sacrificing") nonhuman animals, along with causing them extraordinary levels of pain, debility, and trauma. And all of this just seems to be "common sense."[17]

The research team of Portillo and Paredes has published at least two papers (Portillo and Paredes 2019; Ventura-Aquino and Paredes 2017) in which they place discussions of asexual phenomena in nonhuman animals next to those of asexuality in humans, in a way that is deeply telling. For instance, in an article titled "Motivational Drive in Non-Copulating and Socially Monogamous Mammals," they discuss the concept of "sexual motivation" as it applies to both noncopulating and socially monogamous mammals. In the section about the former, they include a subsection on noncopulating male rats and a subsection on asexuality in humans. The first summarizes the research, including the various interventions that can induce sexual behavior in noncopulating male rats. It also twice utilizes the language of deficit to describe noncopulating male animals. For example, the authors write, "The lower preference for estrous female odors in NC males may be due to *deficits* in the neuronal processing of sexually relevant odors" (Portillo and Paredes 2019, 4; emphasis added). The section on asexuality in humans begins by stating, "The NC males that have been identified in several species could be equivalent to asexual individuals in humans" (5). It then summarizes some of the psychological and social scientific literature on asexuality in humans. The authors conclude by stating, "There is a clear need to understand the biological bases of asexuality. Due to ethical limitations, studies in humans have mainly concentrated on questionnaires and clinical descriptions. However, studies in NC animals suggest that they are present in different species representing a biological variability in which sexual motivation is reduced. More research is needed in this area, not to cure asexuality but to understand and give support to those that could need it" (6–7). Although the authors state explicitly that the goal is not to cure asexuality, ultimately, it is unclear how this research on noncopulating male rats, focused as it is on identifying deficits and methods to induce sexual activity, could provide support to asexual people outside of the logics of cure.

Asexual Possibilities in Scientific Research on
Asexual Phenomena in Nonhuman Animals

While it might be tempting to use the scientific evidence about asexual phenomena in nonhuman animals to argue for the "naturalness" of asexuality in humans, as feminist and queer critics have already asserted in the case of homosexuality, this tactic is limited in significant ways. As we have seen, while evidence for a biological basis of minority sexual orientations can be conscripted to support the "naturalness" of nonnormative sexualities and perhaps encourage people to conceptualize sexual diversity as part of benign natural variation, it can also be used to develop interventions to convert or eliminate minority sexualities. In addition, it can be politically unwise to argue for the equation of "natural" with "benign." Much animal behavior that is "natural" is not something that feminist and queer academics and activists would want to endorse. Most of the evidence produced by scientists for the biological basis of minority sexual and gender orientations is part of a broader research agenda that includes the paradigms of sexual selection theory (evolution produces sex-specific reproductive strategies that are genetically encoded) and brain organization theory (exposure to hormones prenatally hardwires the brain in certain ways, which in turn leads to sex/gender–specific behavior, gender identity, and sexual orientation).[18] Research within this paradigm has provided evidence, for example, for the biological basis of some coercive sexual behavior among both nonhuman animals and humans, of the same type and quality as that produced for the biological basis of minority sexual and gender orientations.[19] We cannot consistently maintain that evidence for the latter validates those identities without implicitly arguing that evidence for the former validates that behavior. We should advocate for the embrace of asexual identities, orientations, and experiences among humans not because they are possibly biologically based but because it is *just* to do so.

This connects as well to another limitation to arguing for the acceptance of asexuality because it is biologically based: I have little doubt that for some (perhaps many) ace people, their asexuality is in part the result of a biologically influenced predisposition, expressed within a particular interpersonal and social context. However, for some (perhaps many) ace people, their asexuality may be a matter of choice and/or the result of a specific "environmental" factor, such as illness, disability, or trauma, particularly sexual trauma. While these experiences of asexuality are excluded by the main AVEN definition of asexuality-as-orientation, they are accepted and embraced by many ace individuals and communities (Barker and Hancock 2019; Kurowicka 2023). Asexuality-by-choice should be embraced because it is just to do so; asexuality due to an environmental factor, including trauma—if the person is content with their asexuality—should also be embraced because it is just to do so.

Although the majority of feminist and queer studies scholars that have written on this issue have focused on critiquing the scientific research on sexuality in nonhuman animals, some have also called for its creative utilization. In general, they assert that studying the sexuality of nonhuman animals may lead us to question some of our commonly held assumptions about human gender and sexuality. For instance, when discussing the relationship between what she calls "transsex phenomena" in human and non-human organisms, Myra Hird (2006, 45) writes, "It is much more interesting to consider how we might understand trans in humans from, say, a bacterial perspective. From such a perspective, given the diversity of sex amongst living matter generally, and the prevalence of transsex more specifically, it does not make sense to continue to debate the authenticity of trans when this debate necessarily relies upon a notion of nature that implicitly excludes trans as a non-human phenomenon." In other words, Hird argues that evidence of transsex phenomena among nonhuman organisms can be used to render "strange" debates about the "naturalness" of the same in humans. Hird also contends that the diversity we find in the sexual behavior of nonhuman animals (including sex for pleasure and to build social bonds) can lead us to question certain cultural assumptions about the "purpose" of sex: "The diversity of sex and sexual behavior amongst (known) species is much greater than human cultural notions typically allow. This diversity confronts cultural ideas about the family, monogamy, fidelity, parental care, heterosexuality, and perhaps most fundamentally, sexual difference" (39).

Related arguments have been made by other feminist science studies scholars, including Stacy Alaimo (2016), Donna Haraway (2003, 2008) and Elizabeth Wilson (2002). For example, in "Eluding Capture: The Science, Culture, and Pleasure of 'Queer' Animals," Alaimo writes that "queer animals" elude "capture" because they defy human categories related to reproduction, sexuality, and gender. Encounters with queer animals (in the flesh or through texts) can provoke wonder and pleasure, she writes: "Wonder may be aroused by that which cannot be understood through simplistic explanations, and pleasure may be inflamed by the sense of being overcome by the staggering variation and the sheer exuberance of more-than-human sexualities and genders" (42), adding, "Queer animal sex may de-sediment intransigent cultural categories" (57). She suggests that not only can encounters with queer animals destabilize mainstream categories of sexuality and gender, they can also challenge and inform feminist and queer understandings of sex and gender. For instance, she states that evidence of four genders among the white-throated sparrow disrupts the feminist contention that gender is a cultural construction. As another, she argues that the fact that most nonhuman animals that engage in same-sex sexual activity also engage in heterosexual activity provides support for "universalizing" models of sexuality as opposed to "minoritizing" ones.[20] Alaimo concludes, "The wonder, awe, and pleasure of contemplating the countless modes of nonhuman sexual diversity, which pulse with desire and erotic ingenuity, may generate environmentalisms that are, of course, already fabulously queer" (62).

Based on the arguments of Hird and Alaimo, I do not claim that evidence of asexual phenomena among nonhuman animals provides evidence for the "naturalness" (and therefore goodness) of asexuality among humans. Rather, I make the somewhat different, although related, claim: that it provides support for the claim that individual variation in sexual motivation is not unusual among at least certain types of nonhuman animals. Thus, the evidence of asexual phenomena in nonhuman animals can render "strange" or "de-sediment" our conviction that the occurrence of the same in humans is something remarkable, something in need of explanation. Refigured in this way, asexual phenomena in humans become mundane, expected, unremarkable—no more in need of explanation than any other form of (a)sexuality.

Toxicity: When Animal Queerness or Aceness Is Caused by Human Activity

In this section, I return to the question of whether ace and queer activists should celebrate evidence of aceness or queerness in nonhuman animals when it appears to be the result of human environmental destruction. Over the past few decades, a great deal of scientific research, popular media coverage, right-wing discourse, and progressive activism has focused on the effect of environmental toxins—most often endocrine disruptors—on human and nonhuman sexuality, gender, and reproduction. Scientists have claimed that environmental toxins have produced same-sex sexual behavior, intersex biology, transsex phenomena, and reduced sexual activity and reproduction in animal species such as frogs and birds. In addition, some research has identified declining average levels of testosterone and average sperm counts among human males, which have also been attributed, in part, to environmental toxins, including endocrine disruptors. As queer and feminist commentators have pointed out, these accounts, along with some environmental campaigns based on them, have often been steeped in heterosexism, identifying any change to sexed embodiment or sexual or reproductive behavior as negative. Right-wing activists and commentators have taken this heterosexism to even greater extremes, identifying the specter of "gay frogs" and the "male fertility crisis" as threats to white, hegemonic, masculine futurity. Although not noted by queer and feminist critics, these scientific, environmental, and right-wing discourses are also often steeped in compulsory sexuality, as one of the "concerning" behaviors identified is decreased sexual activity (R. Lee and Mykitiuk 2018; O'Laughlin 2020; Perret 2020; Pollock 2016; Boast 2022; Kier 2010; M. Murphy 2017; Di Chiro 2010; Seymour 2018). This contemporary concern over the "male fertility crisis" echoes those about race suicide raised by eugenicist sexologists in the late nineteenth century (as discussed in chapter 2).

Some queer and feminist academics have not only challenged the heterosexism, ableism, and racism in these scientific, environmental, and right-wing discourses but have also called for a "reclaiming" of some of the animals altered by environmental toxins, and the embrace or even celebration of queer animals—even when their queerness is the product of technoscientific disruption. For example, in a discussion of the "gay frog," Hannah Boast (2022, 675) argues that, "turned gay" by environmental toxins, it can represent "an antinaturalist account of queerness that acknowledges our immersion in chemical atmospheres while still allowing the potential to advocate for all forms of queer life." Somewhat similarly, Anne Pollock (2016, 183) writes, "I want to suggest that we depathologize queer animals, even when that queerness is the product of human-produced toxins in the environment, and even when it inhibits animals' reproductive capacity. Perhaps we even might find a perverse joy here." Of course, as these scholars recognize, this is a complicated move: Current environmental toxins are themselves the product of exploitative, racist, sexist systems of capitalist production; they can reduce reproductive options for human and nonhuman animals; and their effects are often disproportionately experienced by marginalized human communities, particularly Indigenous, poor, and communities of color, as well as by nonhuman animals who have no say in these matters (Gochfeld and Burger 2011; Johnston and Cushing 2020; Sprinkle and Payne-Sturges 2021).

I believe this may be a situation in which multiple, contradictory truths exist simultaneously. It is wrong to assume, a priori, that changes in the environment that led to disability, altered sexed embodiment, same-sex sexual behavior, and reduced sexual behavior and/or reproduction are necessarily harmful. It is also a problem when efforts to organize against the release of toxic chemicals into the environmental reinforce a kind of reproductive futurism exclusive of certain kinds of queer life. It is also the case that there is no way for humans to live without affecting the environment, which, in turn, affects the shape of human and nonhuman bodies and lives, so we cannot simply assert that it is wrong for humans to alter the lives of nonhuman animals. However, it is yet also the case, generally although not always, that nonconsensual body modification among humans is wrong, as is nonconsensually altering a person's gender or sexuality, particularly when it is done to marginalized individuals and groups, and so we might also see these as harms when done to nonhuman animals (although the concept of consent doesn't translate easily to them). I also generally agree with the tenets of the reproductive justice movement—specifically, the argument that people have the "right to maintain personal bodily autonomy, have children, not have children, and parent the children [they] have in safe and sustainable communities" (SisterSong, n.d.), along with the argument that the release of environmental toxins that reduce options for reproduction, particularly when they disproportionately affect marginalized populations, is a violation of reproductive justice (see, e.g., Liddell and Kington 2021).

This debate over whether we can find pleasure in queer animals produced by environmental toxins that disproportionately affect marginalized human and nonhuman communities is similar in many ways to one in disability studies about the appropriate response to disabilities caused by racism, war, poverty, and other systems of oppression. Academics and activists in this field have long maintained that people with disabilities do not have inherently inferior bodyminds; rather they have ones that differ in some way from the norm, and are stigmatized and excluded through the construction of inaccessible environments (Garland-Thomson 2011).[21] These activists and academics have critiqued a variety of progressive campaigns, including those for environmental justice, that have asserted that pollution, for example, is wrong because it produces disability (see, e.g., Kafer 2013). This is a valuable intervention. Maintaining that something is wrong because it produces disability suggests that disability is something we would like to avoid (eugenic logic), which contributes to stigma against people with disabilities. In addition, as I argued above in the case of asexuality, there may be people who have acquired a disability due to trauma, violence, and/or oppression, who still find value in their disability and in adopting a "disability identity." It is possible that some of the nonhuman animals "turned gay" by environmental toxins are "living well" (and humans might even be able to determine whether this is true on a case-by-case basis through careful observation). However, in *The Right to Maim: Debility, Capacity, Disability*, Jasbir Puar (2017) criticizes disability studies and activism for endorsing a pride model of disability for cases in which the disability is caused by economic and political oppression, which she calls "debility." She contends that insisting on such a pride model reinforces systems of privilege and leaves racist and colonial violence unchecked. Similarly, insisting on a "pride" model for nonhuman animals "turned" queer by environmental toxins might do likewise.

Thus, recognizing these contradictions, I argue that it is an option to find asexual pleasures in "sexually uninterested" giant pandas, altered by human captivity, and in sexually inactive frogs and birds, altered by human-released environmental toxins, but this is not, and cannot be, an innocent pleasure. I argue that it is not a perverse pleasure either, as suggested by Pollock, to the extent that "perverse" is a synonym for "queer"—or at least it is not *just* a perverse pleasure. Rather, it is a morally compromised kind of pleasure.

Rereading Scientific Research on the Sexuality of Nonhuman Animals: What Exactly Is Sexual Behavior Anyway?

Rather than using scientific evidence for asexuality or queer sexuality in nonhuman animals to support ace and queer political projects, I argue here that it may be more productive to use it to problematize our own understanding of the category of "sexual behavior or activity." Rereading scientific studies on the sexuality of nonhuman animals through the analytic of asexuality can "desediment" the cultural category of "sexual activity," as it reveals that its definition in nonhuman animals is no straightforward matter. As we have seen, identifying sexual motivation or desire (or its absence) in nonhuman animals depends on identifying whether they engage in sexual activity, but as Willey suggests in her analysis of studies on voles, "sexual behavior" can be an interpretative grid that human researchers impose on nonhuman animals.

Animal researchers have themselves grappled with how to define and recognize sexual behavior in nonhuman animals. They universally define copulation (specifically defined as a male animal introducing sperm into a female's reproductive tract, often, in mammals, through penile-vaginal penetration) as "sexual behavior." However, most define additional nonreproductive behaviors as sexual, including most that involve an animal (usually a male but in some cases a female) using its genitals to penetrate another animal (either a male or a female, in any orifice), as well as partnered genital stimulation of various kinds (touching, rubbing, grinding, licking, sucking). Animal behaviorists also define solo masturbation in animals as a sexual activity, although it may be harder to identify. In males, it is sometimes easier if the behavior involves erection, genital stimulation, and ejaculation, but, as Roth et al. (2023, 791) note, it may be more difficult to observe in females; as a result, it is sometimes inferred.[22] For example, in his popular book *Biological Exuberance: Animal Homosexuality and Natural Diversity*, Bruce Bagemihl (2000) discusses the case of a female macaque who inserted a variety of tools (leaves, sticks, and other objects) into her vagina. The scientist who observed the behavior hypothesized that the macaque was scratching due to irritation, but Bagemihl reinterpreted the macaque as a creative sex toy user (see also Roth et al. 2023). The point here isn't to adjudicate the "truth" in this particular case but to demonstrate that scientists are engaging in interpretive work when they label an activity "sex."

The need for interpretation when it comes to sexual activity in nonhuman animals is even more apparent in some cases. For instance, in some primates, mothers sometimes initiate "copulation-like" behavior with a male offspring, the purpose of which seems to be to soothe an upset infant. In bonobos, males sometimes engage in "copulation-like" behaviors with other males to reduce conflict or aggression, and this behavior does not necessarily involve sexual arousal for either (Furuichi et al. 2014). Both of these activities might involve genital contact, but are they sex? The answer depends on one's interpretative tools and one's definition of sex. For a long time, animal behaviorists described most nonreproductive "sex-like" behavior as "sociosexual," arguing that the purpose of genital contact between juveniles and adults was preparation for reproductive sex, and that between adults of the same sex, or adults of the opposite sex when reproduction was not possible, was primarily to display dominance (through forced genital contact), to avoid infanticide (females might have sex with a male while pregnant in a bid to prevent him from killing her offspring), to build social bonds, and/or to resolve conflict. However, animal behaviorists (in addition to feminist and queer critics) asserted that this interpretation sometimes or often reflected homophobia (refusing to see same-sex genital contact as sex) or a refusal to see that nonhuman animals might have sex for pleasure, as mentioned above. In turn, many animal behaviorists have reclassified some of these behaviors as "sexual" rather than "sociosexual" and have focused on the motivating role of pleasure (Roth et al. 2023; Furuichi et al. 2014; Clarke et al. 2022; Balcombe 2009). In a 2022 review article, Clarke et al. suggest that scientists should gather more data about sex for pleasure among nonhuman animals by tracking changes in neuromodulator hormones related to experiential pleasure, such as oxytocin. But, even if scientists are able to demonstrate convincingly that this instance of genital contact was to assert dominance, while another was to experience pleasure, would we know if one or both is sex?

The focus on "sex for pleasure" among nonhuman animals leads to another question: Is there a difference between it and other bodily stimulation or contact for pleasure? Animal behaviorists have observed a wide variety of bodily contact—such as hugging, grooming, rubbing, and kissing—that nonhuman animals seem to engage in for many of the same reasons that they sometimes seem to engage in genital contact, including dominance displays, social bonding, and pleasure (Bagemihl 2000). Are these activities sex? If not, why not? Confronting the interpretative work that scientists do in "observing" sexual behavior among nonhuman animals suggests that the category itself is neither self-evident nor necessarily coherent. Thus, rereading scientific findings on animal sexuality through the analytic of asexual possibilities contributes to destabilizing the sexual/nonsexual binary. In the conclusion of this book, I argue that denaturalizing categories, including sexual activity and desire, can foster new ways of talking and thinking about the activities, pleasures, and relationships that we currently classify as "sexual" or "nonsexual," which, in turn, can lead to a more ace-friendly world.

In the next chapter, "Amoebas Are Us? Asexual Reproduction and the Category of Sex," I examine scientific studies on asexual reproduction through the analytics of compulsory sexuality and asexual possibilities. This reading reveals a strong "pro–sexual reproduction bias" in the scientific literature but also identifies instances in which the discourse about (a)sexual reproduction calls compulsory sexuality into question. It also serves to disrupt the category of sexual reproduction and the overarching category of "sex."

CHAPTER 4

Amoebas Are Us?

Asexual Reproduction and the Category of Sex

IN THE EARLY 2000S, THE ASEXUAL VISIBILITY AND EDUCATION NET-work (AVEN) sold an asexual-pride T-shirt with the slogan "Asexuality: Not Just for Amoebas!" (Westphal 2004). Today, an asexual-pride shirt with the slogan "Asexual not Amoeba" can be purchased from Redbubble.com. Together, these shirts express the contradictory positions ace activists and community members have taken regarding the relationship between asexuality as a (human) sexuality identity and a (nonhuman) mode of reproduction. On the one hand, some ace people and communities have embraced the connection between their own sexual disinterest and the asexual reproduction of nonhuman organisms, including amoebas. In 1997, Zoe O'Reilly posted an article online titled "My Life as an Amoeba," in which she identifies as asexual and describes her experiences with asexuality. Other people who identified as asexual responded to it, creating perhaps the first "online asexual community" (Hinderliter 2009). In 2000–2001, a Yahoo group titled "Haven for the Human Amoeba" was formed, providing another early online community for asexual-identified people, although it was soon superseded by AVEN, which was founded around the same time (Hinderliter 2009).

On the other hand, online lists of "myths about asexuality" and "common [negative] responses to asexuality," often include some rejection of the idea that asexuality in humans is like "budding" or asexual reproduction in plants, starfish, or amoebas (see, e.g., King 2021; Allison 2015; "Ace/Aro Mythbusting," n.d.). In her popular 2015 book, *The Invisible Orientation: An Introduction to*

Asexuality, Julie Sondra Decker includes a paragraph listing all of the things that asexuality is *not*, including the statement "We're not amoebas or plants" (3). Thus, unlike in the case of the pandas (and other nonhuman animals) discussed in the previous chapter—in which ace communities were able to find a lighthearted connection and political affirmation in their presumed shared lack of sexual interest or activity—a least a significant segment of ace discourse rejects the association between human asexuality and asexual reproduction in plants and microorganisms. Some of this may be due to the fact that, as Mel Chen (2012) contends, dominant Western thought places "the vegetative" below "the animal" in a hierarchical valuing of entities, while also associating minoritized groups, especially people with disabilities, with the "vegetative." Much of this rejection, however, is likely based on a desire to separate the term *asexuality* as a human sexual identity from its use as a type of reproductive activity engaged in by many types of organisms, but not (or at least not yet) humans.

In a fascinating blog post titled "The Biology Definition of Asexuality May Have More Impact on the Sexual Orientation Than We Think," author Talia (2019) acknowledges that because asexual people chose a term (*asexual*) with a long history in the biological sciences, asexuality as a sexual identity or orientation will continue to be haunted by its previous meanings. She argues that for asexual-identified people who are not interested in having sex or in having children, the association with the term as it is used in the biological sciences may be positive. However, for those who are interested in romantic relationships, experience a sex drive, want to have children, and/or engage in sexual activity for any reason, the association may be invalidating.[1] Talia suggests that asexual-identified people and communities should acknowledge the connections between the meanings of the term in the biological sciences and those in asexual communities, but ultimately so that the ghosts of the former can be exorcised (or unlearned) in order to create space for asexual-identified people in the world.

In this chapter, I examine the concept of "asexuality" in the biological sciences, specifically discussions of asexual reproduction and explanations for the evolution of sex and sexual reproduction. As noted, asexual reproduction differs significantly from the definition of asexuality-as-identity, which focuses on sexual attraction or desire, but still fits within the definition of asexual possibilities. However, I do not advocate for welcoming the ghosts as kin (i.e., claiming that asexual people are amoebas) or for banishing the ghosts (i.e., claiming that asexual people are not amoebas). Rather, I argue that reading the science of asexual reproduction through the analytics of compulsory sexuality and asexual possibilities reveals a strong "pro–sexual reproduction bias" but also instances in which the discourse about (a)sexual reproduction calls compulsory sexuality into question. This chapter also uses compulsory sexuality and asexual possibilities to analyze feminist and queer engagements with bacterial (a)sex and asexual reproduction. In general, such scholars have claimed that bacterial (a)sex and asexual reproduction provide further evidence for the diversity of sex and reproduction in nature, or, in other words, for the "queerness" of nature. This chapter suggests that although this approach has much to recommend it, it can also reflect and reinforce compulsory sexuality. Turning back to the studies on asexual reproduction, I argue alternatively that they can be utilized not as evidence of a queer nature but as a way of demonstrating the incoherency of the category of sexual reproduction and the overarching category of "sex."

Scientific Work on Asexual Reproduction

Here I offer a brief introduction to scientific research on (a)sex, (a)sexual reproduction, and the evolution of sex, with a focus on key terms and concepts, for those unfamiliar with the topic. This section presents the dominant scientific narrative, which I will unpack later in the chapter. However, even a summary suggests that (a)sex and (a)sexual reproduction are complex and varied phenomena that defy clear definitional boundaries (a point that I will return to later).

Scientists themselves do not employ consistent definitions of sex. Still, in this area of research, they most commonly use the term to refer to processes through which genetic material from more than one organism are combined in one organism (other definitions will be discussed at the end of the chapter). Reproduction refers to the production of new organisms. Sex and reproduction can occur independently, or together through sexual reproduction, in which a new organism is formed combining the genetic material of at least two "parent" organisms (Orive 2020).

Using these definitions, for bacteria and other prokaryotes, reproduction and sex are separate activities. Bacteria reproduce asexually, by replicating their DNA and then dividing into two cells, which are usually genetically identical to the original and to each other (for a short overview, see Schmidt 2002). In the 1940s, researchers discovered that bacteria can acquire genetic material from other bacteria, a process referred to as both "lateral gene transfer" and "sex." All bacterial gene transfer is one-directional—a "donor" cell transfers a single genetic fragment to a "recipient" cell, which then incorporates this material into its own genome (recombination). There are three main types of lateral gene transfer: conjugation, in which the donor cell forms a channel into the recipient and injects it with a plasmid (a small circle of genetic material); transduction, in which a virus transfers genetic material from one bacteria cell to another; and transformation, in which a bacteria cell takes up genetic material floating in its environment. (Kohiyama et al. 2003; Narra and Ochman 2006; Redfield 2001).[2]

While some scientists have claimed that recombination generates genetic diversity and/or transfers useful genetic mutations, this does not seem to be its primary purpose. In conjugation and transduction, genetic transfer is an "accidental" side effect of the infectious activities of plasmids and viruses, respectively. In the case of transformation, the cellular machinery may have evolved for nutritive purposes—in other words, to allow the cell to "eat" genetic material from the environment. Many researchers argue that recombination in bacteria is the by-product of DNA repair and replication, as gene repair mechanisms may employ newly introduced DNA to replace damaged DNA (Redfield 2001; Narra and Ochman 2006; Ambur et al. 2016; Moradigaravand and Engelstädter 2013; Michod 1996).[3]

Matters of (a)sex and (a)sexual reproduction are more complicated in the case of eukaryotic organisms (organisms whose cells have a nucleus), a group that includes single-celled protists, fungi, plants, and animals. Eukaryotes can engage in sex without reproduction (lateral gene transfer), reproduction without sex (asexual reproduction), and reproduction with sex (sexual reproduction). Regarding the first, for many years, scientists believed that in most eukaryotes, "sex" (the combination of genes from more than one organism into one organism) only took place as part of sexual reproduction.[4] However, they are increasingly finding evidence of lateral gene transfer in protists, fungi, plants, and animals. These transfers, which are often the result of viruses or other infectious agents, can take place within "kingdoms" or across them (e.g., A. Richardson and Palmer 2007; Xia et al. 2021; Moran and Jarvik

2010; Graham and Davies 2021). Mounting findings about the prevalence and importance of lateral gene transfer have led some observers to reconceptualize the "tree of life" metaphor commonly used in discussions of evolution (see, e.g., Quammen 2018).

Asexual and sexual reproduction are complex and varied phenomena in eukaryotic organisms. Understanding the diversity of these processes requires an understanding of eukaryotic cell division. Eukaryotic cells can divide by *mitosis*, a type of cell division that is like bacterial reproduction. Diploid (2n) eukaryotic cells—which carry two copies of each chromosome, called homologous chromosomes—can also divide by *meiosis*, during which the cell replicates its chromosomes and then divides twice, producing four haploid (n) cells (gametes), which carry one copy of each chromosome. When two gametes from different organisms fuse to create a new organism, this is called sexual reproduction.[5] During meiosis, prior to division, homologous chromosomes sometimes exchange fragments of genetic material, a process called crossing-over or recombination (Hochwagen 2008; Lenormand et al. 2016).

Today, some single-celled protists, including amoebas, only reproduce asexually through mitosis.[6] Some are capable of sexual reproduction through meiosis and fusion. Most types of fungi spend most of their life as haploid (n) organisms (carrying only one copy of each chromosome), reproducing asexually. Most fungi occasionally enter a sexual cycle during which cells of different mating types can fuse.[7] In many types of fungi, only the cytoplasm fuses initially, leading to cells with two or more nuclei before they eventually fuse. As will be discussed later, some scientists consider the initial fusion of cytoplasm to be a type of "parasexuality" (Nieuwenhuis and James 2016; Watkinson et al. 2015).[8]

Plant life cycles generally involve an "alternation of generations" between a diploid (2n) phase (the sporophyte) and a haploid (n) phase (the gametophyte).[9] The sporophyte (2n) produces spores (n) by meiosis. These spores grow into the gametophyte (n), which produces gametes (n) by mitosis. Two gametes (n) fuse to produce a new sporophyte (2n). If the two gametes come from different plants, sexual reproduction has occurred. As will be discussed later, if the gametes come from the same plant, some researchers classify this process as asexual reproduction while others classify it as sexual reproduction (Beck 2010).[10]

Plants can also reproduce asexually in at least two different ways: asexual vegetative reproduction and apomixis. Most plants can engage in vegetative reproduction, which occurs when a portion of a plant is removed from the "parent" and develops into a new plant that is genetically identical to its parent. Some plants have specialized structures—such as rhizomes—to produce vegetative "daughter" plants. Apomixis is a very different process, more similar to parthenogenesis in animals (discussed below). There are many different types of apomixis, but in general, it is the development of an unfertilized gamete into a new individual plant that is genetically identical to the "parent" plant. Apomixis has been identified in almost eighty, or close to 20 percent, of flowering plant families (Hand and Koltunow 2014; Van Dijk 2009; Cruzan 2018).

Except in some cases of parthenogenesis (discussed below), animals live their entire lives as diploid (2n) organisms. During sexual reproduction, specialized organs produce single-celled gametes (n) through meiosis; these can fuse to form a single-celled zygote (2n), which can develop into a new organism through mitosis. If the gametes come from two different individuals, sexual reproduction has occurred (Avise 2008). Some simultaneously hermaphroditic animals (those that produce both sperm and ova at the same time) can self-fertilize (called "selfing"), which again can be classified as asexual or sexual reproduction, depending on the definition used. Examples of simultaneously hermaphroditic animals include some types of worms, barnacles, and mangrove killifish (Jarne and Auld 2006; Avise 2008; Chasnov 2013; Barazandeh et al. 2013). Animals engage in other types of asexual reproduction as well. Many invertebrate animals can reproduce through "agametic cloning," which includes budding, fission, fragmentation, and gemmule formation (internal budding). Invertebrates capable of agametic cloning include some types of sponges, hydras, anemones, and flatworms, among others (Conn 1991; Hiebert et al. 2021).

Some animals can also reproduce parthenogenically, a process that has some similarities to apomixis in plants. In parthenogenesis, an unfertilized egg can develop into a new organism without being fertilized.[11] The new organism can be either diploid (2n) or haploid (n), depending on the process. Invertebrates capable of parthenogenesis include some types of water fleas, rotifers, aphids, stick insects, ants, wasps, and bees. Vertebrates capable of parthenogenesis include some types of lizards (e.g., the whiptail lizard), some types of fish (e.g., the Amazon molly), and some types of amphibians (e.g., the mole salamander).[12] Some parthenogenic animals (e.g., mayflies) are "facultative parthenogens"—meaning that female members of the species can reproduce sexually or asexually at any time. In some parthenogenic species (e.g., ants and bees), fertilized eggs develop into diploid (2n) female organisms (sexual reproduction), while unfertilized eggs develop into haploid (n) male animals (asexual reproduction). Some parthenogenic species, such as aphids, cycle between sexual and asexual reproduction. Others consist solely of asexually reproducing females, who reproduce by parthenogenesis generation after generation (e.g., the whiptail lizard).[13] In some cases, individuals from "nonparthenogenic species" have been found to have reproduced through parthenogenesis; confirmed cases have been found in snakes, Komodo dragons, sharks, turkeys, and other domesticated birds. Scientists estimate that 0.1 percent of vertebrate animal species can reproduce parthenogenically; rates are thought to be much higher for invertebrate animals. Mammals are not able to reproduce parthenogenically under "natural" conditions (MacKenzie 2020; Schön et al. 2009; Avise 2008; Ogawa and Miura 2014; Doums 2021).[14]

As asexual reproduction preceded sexual reproduction historically, researchers have sought to understand why the latter evolved. They argue that it occurred around 1.5 billion years ago in an early ancestor of modern eukaryotes, and has persisted among most eukaryotes, although often in combination with various forms of asexual reproduction. Asexual-only lineages are only a small percent of plant and animal lineages. In addition, scientists claim that these tend to go extinct over time, and so most existing parthenogenic animal lineages, for example, diverted from sexually reproducing ones relatively recently, with a small number of possible exceptions (Schön et al. 2009; Avise 2008; Fontaneto and Barraclough 2015).[15]

Why did eukaryotic sexual reproduction evolve in the first place, and why does it persist over time? Answering these questions has proved challenging for biologists and remains a topic of considerable debate today. In fact, in the 1980s evolutionary biologist Graham Bell (1982, 19) referred to sex as "the queen of problems in evolutionary biology." Early scientific theories, dating from the 1930s, argued that the main purpose of sex is to increase genetic variation and speed evolution. This explanation was accepted for four decades among researchers and remains "conventional wisdom" among laypeople today, present in media articles and high school textbooks (Colegrave 2012, 776; Otto 2008; Meirmans 2009).

However, in the 1970s, evolutionary biologists John Maynard Smith and George C. Williams criticized this explanation for its reliance on the theory of "group selection." They suggested that sexual reproduction is much more "costly" than previously realized—in other words, it significantly reduces the genetic contributions an organism is able to make to the next generation, particularly if a species is sexually dimorphic (with a "male" form that produces one type of gamete and a "female" one that produces a different type) (Meirmans 2009).[16] Maynard Smith offered the following illustration: In a parthenogenic population, all members can produce offspring (say, at a rate of one per year). In a sexually reproducing population, only half of the members (females) can produce offspring (still at a rate of one per year). In this case, the per capita birth rate of the asexual population will be twice that of the sexual population (the twofold cost of sex, or, more accurately, the twofold cost of males) (Lehtonen et al. 2012; Gibson et al. 2017).

As a result of Williams's and Maynard Smith's claims, scientific interest in the "problem of sex" increased significantly. Early mathematical models informed by their arguments demonstrated that sexual reproduction would be "selected for" only under very specific conditions, unlikely to occur often in the real world (Colegrave 2012; Otto 2008).[17] Today, the problem remains "unsolved," but many scientists agree that sexual reproduction evolves in situations in which it is beneficial to break down existing genetic linkages (called linkage disequilibrium), and that a variety of factors may be responsible for creating these conditions (a "pluralistic" explanation) (Colegrave 2012; Otto 2008; Meirmans 2009; Meirmans and Strand 2010; Schön and Martens 2018; Otto and Lenormand 2002). Notably, while there can be sexual reproduction in multicellular eukaryotes without the "cost of males" (for instance, in some types of green algae), researchers have often investigated the two questions (why sexual reproduction? and why males?) together (Lehtonen et al. 2016).[18]

Today, the most accepted theory for why sexual reproduction evolved initially is the need for gene repair (the "repair hypothesis"). As eukaryotes have much larger genomes than prokaryotes, there is a much greater possibility for error or damage during gene replication. Eukaryotes can use homologous chromosomes to repair each other during meiotic cell division. But if both homologues are damaged, sexual reproduction could allow an organism to bring undamaged DNA from another organism into its genome, which can then be employed to repair damaged chromosomes. Evidence for the repair hypothesis includes the fact that some of the cellular machinery responsible for gene editing and repair is the same as that used for meiosis (Tóth et al. 2022; Orive 2020; Michod 1996).[19]

There are two main explanations for why sexual reproduction persists, both of which explain why breaking down existing genetic linkages is advantageous enough to outweigh the costs of sex.[20] According to the first explanation (sometimes called "the Red Queen hypothesis"), sex is needed to promote adaption to an environment that changes over time or space.[21] In particular, the interaction between species (including competition, predation, and parasitism) over time can increase the benefits of sexual reproduction. For example, in a parasitic relationship, a "parasite-resistance" mutation can spread in the host species through sexual reproduction. In turn, a "resistance-overcoming" mutation can spread in the parasite through sexual reproduction. Then, a different "parasite-resistance" mutation can spread in the host, and so on and so forth.[22] The Red Queen hypothesis has been refined over the years and subject to various kinds of testing, with mixed results (Hartfield and Keightley 2012; Orive 2020; Michod 1996; Otto 2008).

The second group of explanations for the maintenance of sex is focused on its ability to remove deleterious mutations through genetic reshuffling. The reality is that most mutations are deleterious rather than beneficial. Since the 1930s, scientists have claimed that one of the purposes of sex is to remove deleterious mutations from a finite population (Hartfield and Keightley 2012; Orive 2020; Michod 1996; Otto and Lenormand 2002). In the 1980s and 1990s, Alexey Kondrashov expanded this theory, arguing that it applies to infinite populations as well, if deleterious mutations act synergistically (if multiple deleterious mutations occurring together create a larger disadvantage than would be expected by adding them together). These theories hold that sexual reproduction allows deleterious mutations to combine in a single individual, and when this individual dies or is unable to reproduce, they are removed from the population. Again, like the Red Queen hypothesis, the mutational deterministic hypothesis has been refined and tested in various ways, with mixed results (Schön and Martens 2018; Hartfield and Keightley 2012).

Critical Scholarship on Scientific Research
on (A)sexual Reproduction

There has been little critical attention to scientific research on the prevalence and types of asexual reproduction among living organisms. Among feminist and queer scholars, this lack of interest is not necessarily surprising, as it does not seem to be immediately related to the status of women or gender-based oppression.[23] The lack of interest in scientific writing on evolutionary theories for sexual reproduction is perhaps more surprising, but most queer and feminist commentators have focused their attention on studies on sexual selection (or evolutionary changes that take place due to intraspecies competition for mates and mate choice). Sexual selection only begins to operate after the development of sexual reproduction and anisogamy (when a species produces two different-sized gametes—male and female), and probably not to a significant extent until the development of gonochorism (when some members of a species only produce male gametes and some only produce female ones).[24]

It does make sense that feminist and queer analysts would focus on studies of sexual selection, as this is the body of research that argues that different levels of "parental investment" by male-gamete-producing organisms ("males") compared to female-gamete-producing ones ("females") led to the evolution of different morphologies, psychologies, and behaviors in males and females (see, e.g., Trivers 1972). Most feminist and queer commentators have been highly critical of this research, contending that it reproduces gendered stereotypes in its association of maleness with competition and sexual aggression and femaleness with nurturance and sexual selectivity (Fausto-Sterling 2008; Hubbard 1990), but some feminist thinkers have maintained that these studies provide information about average differences between genders, as well as gendered biological dispositions, which can inform feminist policies aimed at creating gender justice (see, e.g., Vandermassen 2005).

While not focused on evolutionary explanations for sexual reproduction per se, critical scholars have devoted significant attention to the history of evolutionary biology, elucidating its close relationship to racism, colonialism, and eugenics. In *Ghost Stories for Darwin: The Science of Variation and the Politics of Diversity*, Banu Subramaniam (2014, 7) demonstrates the importance of tracing the history of the concept of variation in evolutionary biology, tied, as it is, to the history of eugenics (political efforts to select for certain types or groups of people and against others): "Questions of genetic variation in human and nonhuman organisms are deeply linked to questions of diversity and difference in human populations steeped in tortured histories of slavery, colonialism, and genocide."

As evolutionary explanations for sexual reproduction are tied to the concept of variation in evolutionary biology, they are thus tied to the history of eugenics.[25] The work of August Weismann, who offered one of the earliest explanations for sexual reproduction, was important for early eugenicists, as his theory of sex allowed them to account for variation while retaining a commitment to genetic determinism (Larson 2010; Sussman 2016; Cowan 2016). Ronald A. Fisher and Herman J. Muller, who offered evolutionary explanations for sexual reproduction in the 1930s (later synthesized as the Fisher-Muller hypothesis), were both committed eugenicists, although Fisher was a political conservative and Muller was a socialist. Evolutionary biologists studying "the problem of sex" in the 1960s and 1970s, including Maynard Smith and Williams, rejected eugenics, but their interest in sexual reproduction was in part motivated by their critique of "group selection" theory (Meirmans 2009). In turn, this provided a foundation for the gene-centered view of evolution, which sees the gene as the fundamental unit of natural selection, and which gained popularity in the 1970s and 1980s (Boldgiv 2010; Ågren 2016). Feminist thinkers have been highly critical of this view of evolution, pointing out that gene-centered explanations have led to a deterministic and inflexible attitude toward sexual difference (Drury 2013; R. Gray 1997).

Pro–Sexual Reproduction Biases in Scientific Research on (A)sex and (A)sexual Reproduction

Here I suggest that the scientific literature on (a)sexual reproduction consistently reflects a pro–sexual reproduction (antiasexual reproduction) bias. However, it is important to keep in mind that this is a large and complex body of scientific literature spanning multiple fields, with over a hundred years of history. Any broad statements about it will leave out disagreements and differences in the literature. It is also important to state, from the beginning, that a pro–sexual reproduction bias is not necessarily equivalent to compulsory sexuality—after all, sexual reproduction includes sex (in the scientific sense of combining genes from at least two sources) and reproduction, while asexual reproduction includes only reproduction (not sex), but asexually reproducing organisms can still engage in sex in the scientific sense through lateral gene transfer. At the end of this section, I argue that although it is difficult to pinpoint a prosex bias in this body of science, its popular circulation does contribute to compulsory sexuality.

In much of the scientific literature, asexual reproduction is associated with "less complexity," since it is primarily bacteria, protists, and fungi that function this way. When it comes to plants and animals, asexually reproducing species are sometimes described as evolutionary "dead ends," because their lineages are thought to have elevated extinction rates (e.g., Takebayashi and Morrell 2001; but see Tripp 2016; Schwander and Crespi 2009). The few asexual lineages that are thought to be ancient (bdelloid rotifers, darwinulid ostracods, and oribatid mites) are described as "evolutionary scandals," and researchers continue to look for evidence of hidden ("cryptic") sex in them (Schwander et al. 2011; Schön et al. 2008; Schwander 2016).

Sexual reproduction, on the other hand, is associated with (and seen as enabling) more "complex" forms of life. Evolutionary biologist and philosopher Richard E. Michod (1996, xix) makes this claim in more poetic language than most, but it still reflects a common belief in the scientific literature when he writes that "sex has enabled life to diversify and expand and, hopefully, to continue to do so forever." More prosaically, Lian Chen and John J. Wiens (2021, 8) write, "Sexual reproduction may be widespread relative to asexual reproduction (in part) because it increases diversification rates among major clades, and possibly because of its association with multicellularity (given that multicellularity has a stronger impact on diversification rates)."

An analysis of the history of explanations for the evolution and maintenance of sexual reproduction is especially revealing of pro–sexual reproduction bias. It could be argued that a recognition of the "costs of sex" and the use of terms such as *paradox* and *problem* to describe sexual reproduction since the 1970s reflects a challenge to prosex biases. It is possible that when students and/or the general public encounter the idea that sex (and males) are costly, this can denaturalize an easy acceptance of sexual reproduction (and sexual dimorphism) as just "the way things are," at least initially.[26] However, a recognition of the high costs of sex (and males) ends up often having the opposite effect—basically, it leads to the conclusion that sex (and males) must confer enormous advantages (even more than previously thought) if they are maintained despite their cost. The idea that sexual reproduction must be highly advantageous is widely (although certainly not universally) expressed in the scientific literature. As just one example, in a review article, geneticist James F. Crow (1994, 205) writes, "Despite the obvious efficiencies of many forms of asexual reproduction, sexual reproduction abounds. . . . From the nineteenth century, it has been recognized that, since there is no obvious advantage to the individuals involved, the advantages of sexual reproduction must be evolutionary. Furthermore, the advantage must be substantial; for example, producing males entails a two-fold cost, compared to dispensing with them and reproducing by parthenogenetic females."

The claim that sexual reproduction produces or increases variation also potentially reveals pro–sexual reproduction biases. As we have seen, this held sway until the 1960s, fell out of favor in the 1970s, and has slowly regained wide acceptance. For instance, Sarah Otto (2009, S11), a leading figure in the field, writes, "So why do so many species reproduce sexually, given the costs of sex and the widespread potential for asexual reproduction? The answer may very well be to reintroduce variation, as suggested by Weismann (1889) more than a century ago." Similarly, Nick Colegrave (2012, 777–78), another researcher in the field, writes, "Thus, it seems that the original intuition of evolutionary biologists was correct after all: the evolutionary success of sex is down to the diversity that it creates. . . . In a complex world in which environments are constantly changing, competitors, parasites and prey are constantly evolving and mutation is continually eroding adaptation, the differences produced by the sexual cycle provide an important evolutionary advantage."

Now it may be technically correct that sexual reproduction is maintained because it creates or increases variation and "diversity," but this is a rosy way of describing it. It is important to note that mutations are the original source of genetic variation; sexual reproduction (merely?) reshuffles genetic material (Avise 2008). As we have seen, two of the major explanations for the maintenance of sex are arguably kind of bleak, depending on your perspective—either sex is maintained because organisms are in a constant battle with other organisms (parasites) seeking to use hosts for their own benefit (the Red Queen parasite hypothesis) and/or so that deleterious mutations can be combined (dumped) into a few individuals who will hopefully die off, thereby removing those mutations from the population (the mutational deterministic hypothesis). It definitely sounds better to say that sexual reproduction is for variation and diversity.

Finally, the lack of research into asexual reproduction and asexually reproducing species may also reflect a pro–sexual reproduction bias. Evolutionary biologist Lynn Margulis has suggested that the scientific research on a(sexual) reproduction reflects a "big like me" bias—we, as humans, reproduce sexually (and most of the organisms we see and knowingly interact with do likewise), so we assume that sexual reproduction is normal (Margulis and Sagan 1990; see also Hird 2009b). Evidence for this argument comes, for example, from a quote offered by a scientist in a news article on his research: "'There is a high cost associated with producing males,' said William R. Rice, first author and professor of biology at the UC, Santa Barbara. 'Just do the math and you will see. And yet, *look out the window and almost every organism you see* reproduces sexually'" (Brown 2001, emphasis added). Contemporary researchers affirm a continuing lack of attention to asexually reproducing organisms. For instance, in their discussion of invertebrate clones, Hiebert et al. (2021, 198) note the lack of scientific attention to these organisms, writing, "Though clonal and colonial animals are a major component of global biodiversity (Jackson 1977), '. . . so much of the study of ecology and evolution has been based on the behavior of unitary organisms (Harper 1985).'"[27]

As noted at the beginning of this section, a pro–sexual reproduction bias is not necessarily equivalent to a prosex bias. In scientific discussions, sexual reproduction encompasses sex (as in the combination of genes from different organisms), reproduction, sex (as in sexual dimorphism or male/female) in sexually dimorphic species, and sexual activity (as in copulation or mating) in species that engage in these activities (the difference between these definitions will be discussed later in the chapter). So, the pro–sexual reproduction bias in the scientific literature may reflect and reinforce a variety of biases other than or in addition to prosex ones.

Yet, even if we cannot isolate the motivations behind the pro–sexual reproduction bias in the literature, I argue that the popular circulation of this research serves to reinforce compulsory (hetero)sexuality. In studies, scientists often use the terms *sex* and *sexual reproduction* interchangeably. When researchers talk about the advantages of sex (meaning sexual reproduction), they are read by others as talking about the advantages of heterosexual activity/mating/copulation. This is most clearly revealed by press releases and news articles reporting on scientific findings in the field, which boast titles such as "Sex and the Single Snail: Study Shows Benefits of Sexual Reproduction over Asexual" (Samarrai 2009) and "Sexual vs. Asexual Reproduction: Scientists Find Sex Wins" (Brown 2001). The first begins with the question "Why have sex?" and then, a few sentences later, asks, "Why not reproduce asexually?," revealing a conflation between "having sex" and sexual reproduction. Other articles are similar. For instance, one titled "A Passion for Survival" begins as follows: "Birds do it, bees do it, and of course, humans. But exactly why we all have sex has been one of the mysteries of science. Now researchers believe they have found the answer" (Fernandez 2016; see also Goodrich 2012; Dockrill 2019). Thus, in the popular press, scientific studies about (a)sexual reproduction become part of the discursive operation of compulsory sexuality, proclaiming the enormous advantages of "having sex."

Some of the scientific research on asexual reproduction does resist the prevalent pro–sexual reproduction bias. In particular, scientists who research asexually reproducing organisms have produced a wealth of information about their complexity and diversity. Hiebert et al. (2021, 198), the researchers studying invertebrate clones mentioned above, point out that "animals that adopt clonal reproduction or colony formation challenge ecological and evolutionary theories often developed to implicitly fit populations of sexually reproducing organisms." A number of scientists who study organisms that reproduce through parthenogenesis or apomixis have found that these clonal lineages express fewer deleterious mutations than predicted by hypotheses about the selective advantage of sexual reproduction. Researchers have begun to identify a variety of mechanisms used by asexually reproducing species to gain some of the benefits of sexual reproduction, including hybridity, occasional sexual reproduction, modified meiosis that allows for some genetic shuffling, lateral gene transfer, kleptogenesis ("stealing" paternal DNA from sympatric sexual species), mitotic recombination, gene conversion (genetic changes that can occur during DNA repair), clonal competition (competition between different clonal lineages, which leads to the survival of better-adapted clones), and epigenetic changes (Verhoeven and Preite 2014; Schön et al. 2008; Mandegar and Otto 2007; Warren et al. 2018; Dalziel et al. 2020; Fradin et al. 2017).

Some researchers have challenged the idea that sexual reproduction is prevalent among multicellular eukaryotes because it confers a major selective advantage. Margulis takes this argument perhaps the furthest, claiming that while meiosis evolved for a reason, "mixis" (as in the fusion of gametes from different individuals) does not confer any kind of evolutionary advantage: "Mixis was never selected for directly. An inordinate amount of data has been collected in attempts to prove the selective advantage of mixis, especially in animals living in unstable environments (Bell 1982). No such conclusion is available from the evidence: neither in constant nor in varying environments can mixis be shown to confer selective advantage over amictic (nonsexual) life cycles" (Margulis and Sagan 1990, 180).[28] Although her research is popular with feminist science studies scholars (as discussed later in the chapter), her claim that mixis does not confer a selective advantage places her in opposition to the majority of scientific research on the topic, and this argument may not be a viable pathway for disrupting pro–sexual reproduction bias.

However, a number of researchers have undermined the narrative that sexual reproduction confers a major selective advantage in other ways. For example, evolutionary biologists Nathan Burke and Russell Bonduriansky (2017) suggest that the true paradox of sex is the fact that not all species combine sexual and asexual reproduction, as doing so would seem to be the most advantageous. Some who study asexually reproducing species have claimed that the rarity of obligate (only) asexuality among eukaryotic species is not the result of selective disadvantage but rather the difficulty of transitioning to asexual reproduction once sexual reproduction has been established. These scientists have elucidated the numerous constraints that prevent the evolution of parthenogenic reproduction from sexual reproduction (Engelstädter 2008; Hojsgaard and Schartl 2021; Galis and van Alphen 2020; Warren et al. 2018). As an example, in discussing their research on the Amazon molly (a parthenogenically reproducing fish), Warren et al. (2018, 674–76) write, "Taken together, we favour a 'rare formation hypothesis' specifying that clonal species might not be rare because of their inferiority to sexual species, but because the genomic combinations that allow successful survival and reproduction are very specific." Scientific findings about the diversity and longevity of asexually reproducing species have been reported in the media, perhaps challenging expectations about sex and reproduction more widely (see, e.g., Grandoni 2023; Bhat 2018; Robitzski 2022; Li 2021). Thus, while the majority of scientific studies on asexual reproduction reflect and reinforce a pro-sexual reproduction bias, some explicitly critique it. Some feminist science studies scholars have employed this later research to denaturalize normative patriarchal understandings of reproduction. In the next section, I analyze feminist engagements with scientific research on (a)sexual reproduction, arguing that while some of this work undermines compulsory sexuality, other aspects may inadvertently bolster it.

Queer and Feminist Scholarship on (A)sexual Reproduction

While there is not a significant body of feminist or queer analysis of asexual reproduction or evolutionary explanations for it, there have been some scholars who have engaged with this research. Elizabeth Grosz's (2011) writing on "Darwinian feminism" is worth mentioning here, due to its strong pro-sexual reproduction bias. Grosz offers a feminist interpretation of Darwin, finding in his writing a useful focus on variation and change, as well as tools for challenging human exceptionalism. However, Grosz also considers sexual reproduction and sexual difference (meaning the evolution of at least two sexes) to be a fundamental aspect of life and identifies sexual selection as the evolutionary force behind creativity, decadence, art, and an appreciation for beauty. This aspect of Grosz's analysis contains logical inconsistencies and, in my estimation, leads to a regressive politics, as others have also suggested (Hird 2012; Schaefer 2021). Here I simply emphasize that in identifying sexual difference as the "engine" of creativity, Grosz simultaneously characterizes asexual reproduction as productive of nothing of interest. For instance, when glossing Luce Irigaray's position on sexual difference (which she endorses), she writes, "Without sexual difference there may be life, *life of the bacterial kind, life that reproduces itself as the same* except for contingency or random accident, except for transcription errors at the genetic level, but there can be no newness, no inherent direction to the future and the unknown" (101; emphasis added).

Some feminist writers and activists have engaged more "positively" with the idea of parthenogenesis. For example, some feminist speculative fiction has imagined societies of only (presumably cisgender) women who are capable of reproducing parthenogenically. The earliest example of this is Charlotte Perkins Gilman's *Herland* (1915).[29] According to Greta Rensenbrink (2010), some lesbian feminists, cultural feminists, and lesbian separatists in the 1970s and 1980s explored parthenogenesis, seeing it as a possible way to reproduce without men and to sustain all-female communities. A few waited for scientists to develop a technological way to achieve parthenogenesis in humans, while others attempted to reproduce parthenogenically on their own. It is beyond the scope of this chapter to evaluate all-female communities (either speculative or real) or parthenogenesis as feminist political strategies, but such an evaluation would need to pay careful attention to the ways in which these strategies are or are not compatible with trans, crip, and racial justice.[30]

The feminist and queer commentators who have engaged more substantially with asexual reproduction generally seem to view it and forms of microbial sex not tied to reproduction (e.g., lateral gene transfer) as potentially queer or feminist forms of sex and reproduction (Parisi 2004; Hird 2009b; Kirksey 2019; Schaefer 2018; 2021; Griffiths 2015). This analysis is related to the feminist and queer scholarship on "queer animals" that I discussed in chapter 3—in fact, both could be considered part of a broader category of scholarship on "queer nonhuman organisms." As discussed in that chapter, the feminist and queer analyses of sexuality in nonhuman animals sometimes suggest that evidence of nonheterosexual sexual practices can be used to challenge human thinking about sexuality. Occasionally, this writing toes the line of claiming that, in these cases, nonhuman animals are engaging in queer sexual practices and/or are themselves queer. When it comes to bacteria and other microbes, feminist and queer critics still argue that microbial "sex" and reproduction can challenge human thinking about sexuality, but they move much further toward identifying microbial "sex" and reproduction as queer or feminist forms of sex and reproduction. Here I discuss the work of the two scholars who have engaged most substantially with asexual reproduction: Luciana Parisi and Myra Hird.

In *Abstract Sex: Philosophy, Bio-Technology and the Mutations of Desire* (2004), Parisi utilizes information from microbiology to elaborate a new understanding of sex as information reproduction and transmission, which she calls "abstract sex." She argues that although most theories focus on heterosexual mating and reproduction as "the" form of (abstract) sex, this is hardly the only one. Drawing on the research of Margulis, Parisi highlights alternative modes of information reproduction and transmission, identifying lateral gene transfer between bacteria ("bacterial sex") as well as endosymbiosis ("hypersex").[31] She also identifies nonsexual forms of reproduction (such as mitotic cell division and the asexual reproduction of mitochondria) as forms of abstract sex. She contends that even in animals (including humans), alternative forms of abstract sex persist, for instance, in the asexual reproduction of mitochondria in our cells (which she calls "cytoplasmic sex") and in parthenogenesis. Parisi asserts that new technologies have enabled additional forms of abstract sex, such as in cloning or synthetic biology (hypersex).

Parisi's analysis has many selling points from an ace-positive perspective, especially the ways in which it seeks to deemphasize the importance of heterosexual reproduction and mating in human thought. However, she makes several additional moves that I do not find politically useful. She identifies bacterial sex, hypersex, and cytoplasmic sex with the feminine and the masochistic, while identifying meiosis, gamete fusion, and heterosexual reproduction and mating with the masculine and the sadistic. In addition, she maintains that bacterial sex, hypersex, and so on engage in or provide resources for a feminist micropolitics or "microfeminine war machine" countering a patriarchal economy of orgasmic pleasure, binary sexes, sexual filiation, genital sex, sexual reproduction, and Oedipal identification (see especially the conclusion). As I elaborate below, I do not see a justification for identifying biological processes as either feminine/masculine or feminist/patriarchal, and even if I were to agree that mitochondrial replication, for example, is a type of micropolitics, it is unclear to me how it could contribute to a program of collective political action.[32]

In her writing, particularly *The Origins of Sociable Life: Evolution After Science Studies* (2009), Myra Hird attempts to build a "microontology" by engaging seriously with microorganisms, especially bacteria. The book contains a chapter titled "Microontologies of Sex," in which she covers some of the same material that I engage with in this chapter, including bacterial nonreproductive sex and asexual reproduction, as well as theories for the evolution and maintenance of sex. In general, like Parisi, Hird seeks to disrupt our focus on heterosexual reproduction and mating through highlighting alternative forms of sex and reproduction in the kingdom Monera (roughly bacteria). For Hird then, bacterial sex and reproduction become yet more examples of the diversity of sex and reproduction in nature. She writes, "We know especially little about bacterial sex and reproduction: yet within Monera, diversity meets its biological and human imaginative limits" (93).

In an earlier article from which "Microontologies of Sex" draws, Hird (2006, 43) cautions against naming particular species as queer, instead encouraging the use of data about the nonhuman world to question human categories of sex and gender. However, in the same article, she writes, "Our remote ancestors [bacteria] continue to promiscuously exchange genes without getting hung up on sexual reproduction. Bacteria are not picky, and will avidly exchange genes with just about any living organism anywhere in the world, including humans." Despite her own warning, in describing bacteria as "promiscuous," "not picky," and "avid" about gene exchange, Hird comes close to describing them as queer and/or as engaging in queer practices. In a small way, this sexualization of bacteria contributes to the invisibilization of asexual possibilities.

In "Queer Love, Gender Bending Bacteria, and Life After the Anthropocene" (2019), Eben Kirksey explicitly describes bacteria as queer. Strangely, he cites Hird's own caution against calling particular organisms "queer." However, he then states that *Wolbachia* (a type of bacteria) perform a variety of "queer tricks," such as converting female hosts to parthenogenic reproduction and transforming male hosts into reproductively viable females. He writes, "Rather than fitting the script of monogamous love . . . the ecological arenas inhabited by *Wolbachia* have become sites of ongoing polyamorous liaisons. . . . *Wolbachia* create kin as they trade genes with other species—blurring the boundaries between the other and the self. Amidst the chaos of jumping genes and rapid evolution, these wily bacteria continue to work as microbiopolitical agents—making some kinds of life live, while letting others die" (209–10). While I see the playful appeal of this argument, I believe it needs further qualification. It is one thing to recognize that microorganisms act in the world independently of human desires, and that these actions have material effects, including, in some cases, political implications for humans. It is also valid to suggest that the actions of microorganisms can encourage us to rethink our own human concepts and can even be utilized as inspiration for human political organizing. However, it is another thing to claim that microorganisms and/ or their actions are, themselves, feminine, feminist, and/or queer. To me, this is far closer to the imposition of human social categories on nonhuman life than it is to the recognition of the agency of nonhuman organisms and matter more broadly.

There are two additional moves made by the queer and feminist scholars who have engaged with asexual reproduction and/or microbial sex that should be further qualified. The first is to attempt to decenter heterosexual reproduction by arguing that most of the living world reproduces asexually. Hird (2012, 224) writes, "Most life on earth is not animal (or plant for that matter); most organisms are not differentiated by sexual difference. All the different kinds of bacteria on earth are estimated to number 5 million-trillion-trillion (5×10^{30}). There are also an estimated 1.5 million species of fungi, many of these reproducing nonsexually through budding, fission, fragmentation, and sporulation. So, most organisms do not reproduce through sexual reproduction." As another example, in "Queer Ecology," Timothy Morton (2010, 276) writes, "Heterosexual reproduction is a late addition to an ocean of asexual division."

It is certainly true that bacteria and many fungi reproduce asexually. It is also true that even humans (and other animals) reproduce asexually at various levels (DNA replication, mitochondrial replication, somatic mitosis) apart from that of the organism. As scientist John C. Avise (2008, 1) writes, "A clone can be defined as a genetic copy . . . of a previously existing biological entity. Such an entity can be interpreted broadly to include a particular stretch of DNA (a locus), an ensemble of physically linked loci (such as the genes comprising mitochondrial DNA), a genome (the entire suite of DNA) of a somatic cell, or the full genetic constitution of a multicellular organism. In vertebrates, clonal replication occurs universally at the first three of these levels."

However, it is also true that, at the level of the organism, humans, almost all other animals, and many plants reproduce sexually. The following statement by Otto (2009, S1) is also true: "In the face of such legendary costs, we might expect sexual reproduction to be rare. Yet, the vast majority of eukaryotic organisms reproduce sexually—at least occasionally. Among named animal species, only ~0.1% are considered to be exclusively asexual (Vrijenhoek 1998). While the ability to reproduce asexually is more widespread in plants, less than 1 percent of the approximately 250,000 angiosperm species are thought to be substantially asexual (Asker and Jerling 1992; Whitton et al. 2008)." Thus, the claim that asexual reproduction is more common than sexual reproduction may be true, while the claim that sexual reproduction is more common (or even necessary) for multicellular eukaryotic organisms may also be true. Whether one truth is more important than the other is a matter of perspective, not ontology. While it may accomplish something, politically, to remind ourselves that sexual reproduction is not the only, or even not the most prevalent, form of reproduction, if our social justice projects are at all focused on the transformation of human societies, then we must contend with sexual reproduction.[33]

The second move that needs further qualification is the attempt to decenter heterosexual reproduction by arguing that sexual reproduction is not evolutionarily advantageous. As noted earlier in the chapter, Margulis claimed that "mixis" (or sexual reproduction) does not confer any kind of evolutionary advantage, or, in other words, was not selected for directly. Because Parisi, Hird, and other feminist and queer critics who engage with this topic rely heavily or solely on the work of Margulis, they repeat her assertion that sexual reproduction does not confer an evolutionary advantage. For instance, glossing Margulis, Parisi (2004, 69–71) writes, "The entanglement of sex with reproduction (sexual mating), far from being the fruit of selection, is linked to the meiotic process of cellular doubling and reduction. . . . Genetic variations and sexual diversity are not the result of sexual mating as often maintained by evolutionary biologists." Similarly, also glossing Margulis, Hird (2009b, 96) writes, "From this perspective, the kind of sex practiced by some animals (including humans) is a result of the failure of organisms to exchange DNA through other means. As such, sexual reproduction is an outcome of contingent circumstances rather than evidence of greater 'complexity' or hybrid vigor."

Again, there is value to this move. Suggesting that sexual reproduction (mixis) provides no (necessary) advantage offers the possibility that we could be otherwise.[34] However, even Margulis herself contends that although mixis might not have been selected for, almost all animals (including humans) are basically stuck with it. So, as in the first move, this one doesn't seem to take us very far politically. Moreover, as stated earlier, Margulis's argument is contrary to the majority view of evolutionary biologists, which is that sexual reproduction is maintained by selection (although, as noted before, many scientists agree with Margulis that once sexual reproduction has evolved, there may not be many pathways for reverting back to asexual reproduction) (Engelstädter 2008). We need to be prepared politically for the possibility that at least some of the complexity, longevity, and ubiquity of multicellular life is due to sexual reproduction.[35]

Asexual Reproduction: Querying the Human Category of Sex

Although I am sympathetic to the desire of scholars such as Parisi, Hird, and Kirksey to use scientific evidence of bacterial "sex" and reproduction to decenter heterosexual mating and reproduction by enlarging the category of sex and reproduction, here I take a somewhat different approach. Rather than arguing, for example, that bacterial sex is evidence of the diversity of sex in nature, I posit that we can utilize the scientific evidence about (a)sexual reproduction and bacterial "sex" to demonstrate the incoherency, and perhaps limitations, of the category of sex itself.

First, I want to slow down the immediate acceptance of lateral gene transfer as "sex." Lateral gene transfer has little in common with most lay definitions of the term (as in sexual activity). Lateral gene transfer involves neither reproduction nor pleasure. As mentioned, there are three general types in bacteria: conjugation, transduction, and transformation. In the cases of conjugation and transduction, it is aconsensual, and may only benefit the plasmids or viruses involved, not the bacteria. In all cases, including transformation (when a bacterium ingests free-floating genetic material from its environment), it is a nonreciprocal, one-way activity, from donor to recipient. Transformation seems to bear as much resemblance to eating as to sex, as the terms are commonly used (Redfield 2001). It is only if we accept a scientific definition of sex as the combination of genetic material from two different organisms into one that we would call lateral gene transfer sex. However, if we accept this scientific definition, much or most of what laypeople consider sex would not count, as human sex only meets this definition when it is reproductive (as that is the only time when it brings genetic material from more than one organism together into one).[36]

What is the value in calling lateral gene transfer sex? What do we gain from sexualizing bacteria? Feminist science studies scholar Londa Schiebinger (1993) demonstrated that in the case of plants, scientific efforts to conceptualize plants as engaging in sexual activity were informed by and served to reinforce both conservative and progressive gender and sexual ideologies. There is nothing inherently progressive or liberating about sexualizing nonhuman organisms. I suspect that it is only compulsory sexuality that leads us to so easily subsume such disparate activities as lateral gene transfer among bacteria and various forms of sexual activity among humans under the category of "sex." Michel Foucault (1990, 154) argues that the historical invention of sex as a concept "made it possible to group together, in an artificial unity, anatomical elements, biological functions, conducts, sensations, and pleasures, and it enabled one to make use of this fictitious unity as a causal principle, an omnipresent meaning, a secret to be discovered everywhere: sex was thus able to function as a unique signifier and as a universal signified." Is it possible that in "discovering" the existence of bacterial sex, we have merely extended the deployment of sexuality to cover yet another disparate element?

It is one political strategy to expand or proliferate a category as a way of challenging it—this strategy has been proposed in the case of gender, for instance, as some feminists have suggested that a proliferation of categories and identities will help to dismantle the regulatory power of the gender binary and gender inequality (Bem 1995; Lorber 2021). It is certainly possible that proliferating the category of sex, by identifying lateral gene transfer, endosymbiosis, and genetic engineering (among other things) as sex (as Parisi does), may challenge its regulatory power. However, here I pursue an alternate strategy by arguing that the category is neither coherent nor particularly useful.

Rereading the scientific evidence about asexual reproduction introduced at the beginning of this chapter, I contend that it suggests that sex is not a natural category, either for scientists or for the nonhuman living world. Researchers agree that the definitions they use are not consistent. Maria E. Orive (2020, 273) notes that biologists use the term *sex* to describe at least four different phenomena: "(1) the existence of separate sexes (or dioecy); (2) anisogamy, or the fusion of two dissimilar gametes; (3) meiosis, a specialized form of cell division leading to the production of gametes, which may or may not include genetic recombination; and, finally (4) genetic recombination itself." It is worth noting here that scientists employ different definitions of recombination as well, sometimes only meaning the exchange of genetic fragments between homologous chromosomes that sometimes occurs during meiosis (called crossing-over) and sometimes to mean crossing-over plus the random sorting of genetic material that occurs during meiosis when the material is divided between "daughter" cells (independent assortment).[37] As stated at the beginning of the chapter, most researchers use definitions close to the following: sex is the bringing together of genetic material from at least two organisms into one, and sexual reproduction is the bringing together of genetic material from at least two organisms into a new one. However, contradictions appear almost immediately.

Scientists disagree about whether lateral gene transfer among bacteria should be called "sex." Some only use the term to refer to the meiotic division and fusion that occurs in eukaryotic organisms (e.g., Schön and Martens 2018; Billiard et al. 2012). Even when it comes to eukaryotes, there is disagreement about what counts as sexual or asexual reproduction. Many scientists insist that the former requires the combining of genetic material from at least two organisms; for them, self-fertilizing (selfing) hermaphrodites are engaging in asexual reproduction (e.g., Tripp 2016), while others consider them to be engaging in sexual reproduction (e.g., Billiard et al. 2012), while still others separate asexual reproduction, selfing, and outcrossing into separate categories (e.g., Glémin et al. 2019). Most of the scientific studies I reviewed considered parthenogenesis to be an asexual form of reproduction, but there were exceptions. Some forms of parthenogenesis occur through mitosis, in which case there is not genetic recombination (apomictic parthenogenesis), while others occur through modified meiosis, in which case there may be (de Meeûs et al. 2007). Some researchers consider the latter to be a form of sexual reproduction. For example, Billiard et al. (2012) define sexual reproduction as any reproduction that involves a diploid (2n) phase followed by division to a haploid (n) phase followed by fusion to a diploid (2n) phase. Under this definition, those forms of parthenogenesis that involve meiosis would be considered sexual reproduction (see especially figure 1).[38]

In addition to these definitional differences, there are also a number of processes that exist in the living world that blur the boundaries between sexual vs. asexual reproduction, as recognized by scientists, who sometimes refer to these as "parasexual." As already noted, a number of species combine sexual and asexual reproduction; for instance, in many different types of fungi, when gametes fuse, only the cytoplasm fuses initially, not the nuclei, leading to cells with two or more nuclei each (called di- or polykaryotic). Fungi vary in terms of how much time they spend in this state (Nieuwenhuis and James 2016). In some species of parthenogenically reproducing vertebrates, there exists a parasexual process called hybridgenesis, in which the diploid (2n) female employs a special type of meiosis to produce a haploid (n) egg that only contains genes from its own "mother," not its "father." This egg is fertilized by a sperm to produce a diploid (2n) zygote, which grows into a diploid (2n) female organism. When mature, the new organism can again produce an egg containing only genes from its "mother." This form of reproduction could be classified as either sexual or asexual, depending on the definition used (Avise 2008). Or consider the case of inbreeding: If the "point" of sexual reproduction is to bring together material from genetically different organisms, then sexual reproduction between closely related organisms could be considered "less sexual" than that between distantly related ones; in fact, one could maintain that sexual reproduction exists on a continuum from selfing to reproduction between the two most genetically different individuals in a population (for a discussion of inbreeding as a form of asexual reproduction, see Avise 2008).

Arguably, the definition of sexual reproduction as the combining of genetic material from at least two organisms into a new one is somewhat arbitrary. If the selective difference between asexual and sexual reproduction is due to the benefits of genetic recombination (as many researchers claim), it would make more sense to differentiate between different types of reproduction based on some measure of the amount, kind, or level of recombination involved—but different kinds of asexual reproduction (e.g., mitotic or apomictic parthenogenesis vs. meiotic parthenogenesis) and different kinds of sexual reproduction (e.g., inbreeding vs. breeding between distantly related organisms) involve different amounts, levels, and types of recombination.[39] The binary distinction between asexual vs. sexual reproduction fails to capture the diversity of reproduction either in terms of form or function.

The situation becomes even more interesting if one tries to identify not just when sex is occurring, as defined by scientists, but also which organisms can be said to be "having sex" in any sexual event or process (if, indeed, it even makes sense to talk about organisms "having sex" here). If we describe lateral gene transfer as sex (whether among bacteria or other organisms), then we might identify the donor and recipient organisms as "having sex." But, in transformation, as the recipient bacteria takes up genetic material from the surrounding environment, would we consider the donor bacteria to have "had sex"? If the human creation of a transgenic organism in the lab is described as a sexual process, would we describe the donor and recipient organisms as having "had sex"? What about the scientists and lab workers who perform the genetic engineering? In sexual reproduction, the process occurs over more than one generation—the parents produce gametes, and the gametes fuse to produce a new organism—and it is only at that point that sex (as in the combining of genetic material from two organisms into one) has occurred, so would we describe the parents and the offspring as having had sex (together)?

And, of course, this is only one of the broad ways in which "sex" is used in science—as in, the bringing together of genetic material from multiple sources, either with reproduction or without. The second way (as noted by Orive above) is in terms of sexual difference—that is, different types of gametes produced by one species (if there are two sizes of gametes, the smaller is named male and the larger female).[40] Then, in gonochorism, the male-gamete-producing organisms are considered "male" and the female-gamete-producing ones are considered "female." According to Schön and Martens (2018), this second usage of sex "includes all aspects of behavior, morphology, and physiology that deal with gender differences and mate selection which result from sexual selection."

The third broad way in which the term *sex* is used is in the sense of "sexual activity" or "having sex." Most biologists use the terms *sexual intercourse*, *copulation*, *mating*, and/or *coitus* to describe the process by which gametes from a male organism are deposited in the body of a female one during sexual reproduction (R. Hine 2019), but many scientists employ the term *sexual behavior* to describe a variety of nonhuman animal behaviors that "look like" copulation in some way (Dixson 2012). As I have argued elsewhere (Gupta 2022), there is not a single definition that can capture all of the activities commonly described as sex by people—sexual activity can be defined neither by function (people have sex for all sorts of reasons, including reproduction, pleasure, money, intimacy, novelty, etc.) nor by form (all sorts of different activities can be experienced as sex).

My argument here is that each one of the three main uses of the term *sex* contains within it incoherencies. Considering all three together, it is clear that the term is applied widely to different activities and phenomena that bear some resemblance to heterosexual sexual reproduction or mating. Its use serves to obscure the differences between these activities and phenomena while at the same time interpellating them as sexual and constituting them as part of the heterosexual matrix. Sex then appears to be everywhere, always in the human and nonhuman living worlds (even bacteria have sex!).

Rather than suggesting that asexually identified people are or are not like asexually reproducing organisms, my argument is that scientific evidence about (a)sexual reproduction demonstrates the incoherence of the category of sex itself. I contend that we can utilize this research to undermine our own easy acceptance of sex as a "thing and/or activity existing in the world." We might want to stop using the term *sex* to describe and unify so many disparate elements. In fact, we may wish to stop relying on it at all. What explanatory value does it have? Does it accomplish anything other than interpreting the world through the lens of the heterosexual matrix? Perhaps we should try to talk differently about what we (and nonhuman organisms like bacteria) do together and apart. Perhaps this strategy, instead of or in addition to proliferation, might serve to undermine the regulatory power of compulsory (hetero)sexuality and, by extension, imbricated systems of oppression based on gender, race, ability, and other social categories. What variegated language could we come up with to describe moments of pleasure, intimacy, bodily interaction, information exchange, reproduction, splitting, creation, copying, engulfing, and more if we are not calling all of these "sex" (or "asex")? What different ways of thinking and being or becoming might be unlocked by this new language?

Conclusion

Toward an Antinaturalist, Intersectional Ace Politics

DENATURALIZING TAKEN-FOR-GRANTED CONCEPTS IS A HALLMARK OF critical scholarship. Feminist and queer academics have adeptly destabilized understandings of sex (as in male/female), gender, sexual identity, and sexual orientation, including in hegemonic science. Feminist and queer critics have also underscored the ways in which sexual desires and practices are shaped by social norms related to gender, sexuality, race, disability, and more. However, in popular culture, in some feminist and LGBTQ+ activism, and in some feminist and queer scholarship, the concepts of sexual desire and activity (as in "having sex") are often treated as self-evident and self-explanatory. Yet, as the previous chapters have suggested, reading scientific studies on sexuality through the analytics of compulsory sexuality and asexual possibilities reveals that these concepts are characterized by incoherence.

In this conclusion, I briefly review some of the ways in which sexual desire and activity are essentialized in feminist and queer academic work, in asexual activism itself, and in scientific research on sexuality. I contend that understandings of sexual desire as an essential drive reinforce compulsory sexuality, as they position asexuality as deficit and/or dysfunction. I then return to the analyses of research performed in chapters 1 through 4 to argue that examining this research on sexuality through the analytics of compulsory sexuality and asexual possibilities denaturalizes the concepts of sexual desire and activity, demonstrating that their coherence depends on a reference to heterosexual reproduction. I conclude by suggesting that rather than endlessly expanding the categories of sexual desire and activity, we might instead develop new ways of talking and thinking about our desires, pleasures, activities, and relationships that do not consider whether they are "sexual" or "nonsexual" to be their most important characteristic. This move can help to create a more ace-friendly world, which, simultaneously, is one that supports a greater diversity of human flourishing.

Essentializing Sexual Desire and Activity: Everyone Does It

On the one hand, denaturalizing sexual desire has been at the core of much US feminist and queer scholarship since the 1960s. The former, arising out of the women's liberation movement, held that women's heterosexual desire for men was not natural but rather compelled by patriarchal systems of oppression (D. Richardson 1997; S. Jackson 1996). In *The History of Sexuality Volume 1* (1990)—one of the founding texts of queer theory and sexuality studies—Michel Foucault denaturalizes sexual desire, identifying it as part of the broader deployment of sexuality as an apparatus of power: "By creating the imaginary element that is 'sex,' the deployment of sexuality established one of its most essential internal operating principles: the desire for sex—the desire to have it, to have access to it, to discover it, to liberate it, to articulate it in discourse, to formulate it in truth. It constituted 'sex' itself as something desirable. And it is this desirability of sex that attaches each one of us to the injunction to know it, to reveal its law and power" (156–57). For this reason, Foucault states that "the rallying point for the counterattack against the deployment of sexuality ought not to be sex-desire, but bodies and pleasures" (157).[1] Building on as well as departing from the work of Foucault and early feminist scholarship, since the 1980s, feminist and queer analysts have illuminated the ways in which sexual desire is shaped by racism and ableism and incited by neoliberal capitalism (e.g., Hammonds 1996; Shildrick 2007; Hennessy 2000).

However, despite this substantive history of denaturalizing sexual desire, the concepts of "sexual desire" and "sexual activity" are also sometimes taken for granted as self-evident in some feminist and queer academic work and activism. Looking back at US feminist scholarship and activism from the 1970s and 1980s, Jana Sawicki (1991) argues that both radical or cultural and libertarian or prosex feminists tended to essentialize sexual desire. Some radical and cultural feminists viewed male and female sexual desire as essentially different and sought to free the latter from patriarchal oppression. Libertarian or prosex feminists tended not to view male and female sexual desire as essentially different, but some saw the latter as more repressed by society and sought to liberate it. Sawicki (2010) also suggests that a number of queer theorists in the 1990s were hesitant to embrace Foucault's rejection of sexual desire, instead choosing to incorporate psychoanalytic discourses of desire into their work, whether Freudian or Lacanian.[2]

Jane Ward's book *Not Gay: Sex Between Straight White Men* (2015) offers an example of feminist and queer scholarship that (inadvertently) reifies sexual activity. She claims that heterosexual white men engage in significant amounts of same-sex sexual activity. To support this, she analyzes hazing rituals that involve genital contact between fraternity members. Even though many of the participants do not define or experience them as sexual acts, Ward identifies them as "sex" because they involve genital contact and because a "hypothetical queer couple" would identify them as such, suggesting that what counts as sex can be identified by outside observers, regardless of the subjective meaning of the acts to participants. In turn, this suggests that "sexual activity" is a category with some kind of objective truth.[3]

Perhaps ironically, while mainstream contemporary asexual communities have, in many ways, denaturalized sexual desire, in other ways they have essentialized it. As other critics have noted (Scherrer 2008), in asserting that asexual people do not experience sexual attraction, asexual discourse implicitly suggests that it is a self-evident phenomenon. As discussed in the introduction of this book, mainstream asexual groups have also differentiated between "romantic" and "sexual" attraction. Again, while this in some ways destabilizes commonplace understandings of sexuality and relationality, it simultaneously "thingifies" both sex (and sexual attraction) and romance (and romantic attraction)—in this system, sex and desire can be identified and separated from romance and desire, and one can possess the capacity for one, both, or neither. Contemporary asexual and aromantic (aspec) communities have attempted to acknowledge and account for the blurriness of these categories through the creation of terms such as *gray asexual* and *gray aromantic*, which seek to capture those people who fall "in between" sexuality/asexuality and romanticism/aromanticism. But, arguably, a system with more than two categories is still only a marginal improvement over one with two. In recent years, asexual and aromantic discourse has increasingly embraced "spectrum" language—for example, according to these discourses, a person can be on "the" asexual spectrum and/or "the" aromantic one. This embrace of spectrum language is, again, perhaps an improvement over categorical language, but even in keeping with spectrum logics, sexual attraction is separated from romantic, sensual, or aesthetic attraction.[4]

If, at times, feminist, queer, and asexual scholarship and/or activism works to essentialize sexual desire, it is perhaps not surprising that much hegemonic scientific research on sexuality does as well. As has been maintained throughout this book, much of it treats sexual desire and activity as self-evident concepts. Sexual desire is frequently understood (even if implicitly) as an innate drive. As was demonstrated in chapter 1, European and American medical and mental health professionals since the late nineteenth century have often understood sexual desire to be a natural biological drive with a functional purpose (whether reproduction, pleasure, relationality, and/or racial "hygiene") and have therefore often viewed sexual disinterest as a deficit in need of diagnosis and treatment. As discussed in chapter 2, psychologists and neuroscientists using imaging technology to study the neural aspects of sexual behavior, arousal, and desire have officially rejected a drive model of sexual desire, but in their continuing efforts to isolate the unique (as in different from other cognitive and emotional states) but universal (as in the same for all people) neural pathways associated with it, they reflect and reinforce the reification of sexual desire. As was demonstrated in chapter 3, researchers studying the sexuality of nonhuman animals often stigmatize sexually inactive animals and, in some cases, seek to produce sexual behavior in them through the administration of hormones and other drug interventions. In addition, as was argued in these chapters, research on the neurological aspects of sexuality and on sexually inactive animals is often tied to efforts to define and develop "treatments" for sexual disinterest in humans, particularly women.

Although chapter 4, concerned as it is with asexual vs. sexual reproduction, does not analyze the operationalization of sexual desire in scientific studies, it does demonstrate that researchers maintain a binary separation between sexual and asexual reproduction, despite evidence of phenomena that exist on the border between the two concepts. It is also clear that scientists started with an understanding of sexual reproduction and activity (as in mating or copulation) based on the human (or at least mammalian) model, in which it occurs only through this method. They then applied this model and its associated language to describe the reproductive *and* nonreproductive genetic-mixing activities of other types of organisms, including plants, fungi, and, most recently, bacteria, which do not necessarily involve mating or copulation. This model and its associated language were also applied to human (and animal) activities that "look like" copulation but do not involve reproduction or genetic mixing (e.g., sex for pleasure or social bonding). Although there are significant differences in the activities that are labeled "sexual" between these types of organisms (a point that will be returned to below), the use of the same model and language to describe them suggests that bacterial "sex" (in which bacteria may take up free-floating genetic material from their environment) is the same as plant "sex" (in which an insect may carry pollen from one flower to another, fertilizing the second) is the same as "sex" among animals (in which a mule deer may masturbate or two humans may engage in oral sex, one primarily for pleasure and the other primarily for money).[5] Sex then appears to be everywhere, always in the human and nonhuman living worlds.

This essentialization of sexual desire and activity has real-world consequences for aspec people. If desire is conceptualized as a universal, innate drive designed to promote reproduction, pleasure, and/or relationality, and sexual activity as an activity engaged in by all living organisms, then asexuality can only primarily be understood as lack—the lack of a universal and primary motivating drive, the lack of interest in a universally "interesting" activity. As we have seen, this conceptualization of sexual desire and activity undergirds a view of asexuality as either pretense (sexuality will be discovered underneath asexuality eventually), deficiency (asexuality is less than sexuality), and/or dysfunction (asexuality is a distortion of sexuality that can or should be treated). Thus, in many ways, scientific research has both reflected and reinforced compulsory sexuality.

Denaturalizing Sexual Desire and Activity

Although sexual desire and activity are often essentialized in hegemonic scientific research on sexuality, rereading this research through the analytic of asexual possibilities can de-essentialize the intransigent colonial, racist, heterosexist, classist, and patriarchal categories of sexual activity, pleasure, and desire. As discussed in chapter 4, rereading scientific studies on sexual and asexual reproduction denaturalizes the category of "sex." As mentioned earlier, the term has been extended in two directions from the starting point of animal reproduction, in which sex (as in the combination of genetic material from two different organisms) occurs only with reproduction (the production of a new organism) and only with sexual activity, mating, or copulation (the depositing of gametes from a "male" into the reproductive tract of a "female"). Moving in one direction, it was applied to activities leading to sexual reproduction (genetic recombination and reproduction) in plants and fungi (even though these activities look very little like animal copulation) and further applied to activities leading to sex (as in genetic recombination) in bacteria, even though these activities do not involve reproduction and often look very little like copulation. In the other direction, the language of sexual activity was extended to those in animals that "look like copulation" but do not involve sexual reproduction (either genetic recombination or reproduction). This extension of language and concepts has served to unite disparate activities under the colonial category of sex, but rereading this research with attention to these disparities allows us to recognize incoherences.

As was discussed in chapter 3, rereading the studies on sexual activity in nonhuman animals reveals that its identification requires interpretative work. Researchers have grappled with how to define and recognize sexual behavior in nonhuman animals. Copulation is universally understood to be a sexual activity, but some scientists define other activities involving genital penetration or stimulation as "sexual behaviors" as well. But they have also debated whether the purpose of the activity matters—is it "sexual" if it is for reproduction or pleasure? Is it "sociosexual" (rather than "sexual") if it is for dominance, protection, social bonding, and/or conflict resolution? If genital activity is "sexual" if it is for reproduction or pleasure, then what about other, nongenital bodily activities engaged in by nonhuman animals (e.g., kissing, hugging, massaging) for some of the same reasons (Roth et al. 2023; Furuichi et al. 2014; Clarke et al. 2022; Balcombe 2009)?[6]

As was argued in chapter 2, rereading the neuroscientific literature on sexuality denaturalizes the categories of sexual pleasure and desire.[7] Rereading this research through the analytic of asexual possibilities suggests that, perhaps, sexual pleasure and desire are not nearly as distinct from other pleasures and emotional and motivational states as is commonly thought in mainstream US society, and that sexual desire is not as intrinsic as suggested by the drive model of sexuality. Many neuroscientists claim that evolution tied sex (specifically genital stimulation) to a common pleasure/reward system. From an ace perspective then, genital stimulation becomes simply one among a number of activities that produce pleasure, primarily distinguished by the connection to reproduction, the specific body parts involved, and (for humans and some other animals) the social meanings attached. If humans experience some types of pleasurable bodily stimulation or contact beyond the genital as sexual but not others, then this difference between "sexual" and "not sexual" is constructed, not essential. Similarly, the neuroimaging data suggest that "sexual motivation" is similar to other motivational states and feelings. Again, from an ace perspective, it becomes simply one among a number of motivational states, again made "special" primarily by the social meanings attached.

Thus, rereading scientific research on sexuality through the analytic of asexual possibilities helps to unsettle essentialized concepts, including sexual arousal, activity, desire, pleasure, and reproduction. As argued above, essentialized understandings of these concepts contribute to a view of asexuality and other nonsexualities as pretense, deficiency, and/or dysfunction. Thus, critical engagements with science and the material world it seeks to capture can play an important role in challenging problematic ways of thinking, thereby clearing the way for more ace-positive conceptual frameworks.

While I believe the most important role for science in ace political projects is to provide the mental grit that can allow theorists and activists to think differently, research can, in some cases, be used pragmatically to construct new ways of thinking that are more conducive to ace liberation. For instance, I contend that combining an incentive motivation model of sexual response with a developmental systems approach can provide a more ace-positive model for thinking about sexual activity, pleasure, and desire (Gupta 2022).

As was discussed in chapter 2, the neuroimaging research on human sexual arousal, response, and desire provides support for an incentive motivation model of sexual desire. To review, this model views sexual motivation as "the result of the activation of a sensitive sexual response system by sexually competent stimuli that are present in the environment" (Both et al. 2007, 329). For humans, at least, the stimuli may be external or internal, including fantasies. The incentive motivation model of sexual response views sexual motivation as similar to other motivational states or action tendencies, including hunger, thirst, and drug craving (Toates 2009).

Christopher Harshaw's (2008) developmental systems that work on hunger can serve as an additional resource for an alternative model of (a)sexual development. In "Alimentary Epigenetics: A Developmental Psychobiological Systems View of the Perception of Hunger, Thirst and Satiety," Harshaw challenges the idea that the perception of hunger, thirst, and satiety (alimentary interoception) is an "innate drive." According to him, alimentary interoception is the product of individual development. He uses evidence from the study of humans and nonhuman animals to demonstrate that a human infant does not have an inborn connection between specific interoceptive sensations and the concepts of hunger and thirst, between these sensations and specific nutritive environmental stimuli, or between these sensations and the behaviors involved in eating or drinking. Rather, these connections must "be acquired during ontogeny through processes of socialization or social biofeedback," specifically through reciprocal interactions with caregivers (560).

Combining the incentive motivation model with a development system approach like that offered by Harshaw can lead to a more ace-positive model for how to think about the processes through which activities are defined (or not) as "sexual" and those through which sexual pleasure and attraction become stabilized (or not) as a recognizable phenomenon for the individual. Translated into the realm of (a)sexuality, we can think about the individual development or nondevelopment of sexual attraction as an iterative process in which bodily and mental processes intra-act with social norms and systems (including intersecting systems of oppression), as well as socially developed conceptions of sexuality, primarily via ongoing reciprocal interactions with others. Over time, certain activities are defined as sexual, and sexual attraction and sensations are stabilized (or not) as recognizable or distinct phenomena for the person. In addition, over time, through learning, stimuli recognized as "sexual" may increase or decrease in incentive motivation value for a person. This way of thinking about sexuality is useful for thinking about asexuality, because it provides a framework in which the latter is not a lack but rather a different organization of bodies and pleasures—for example, some sensations or feelings that others characterize as sexual may not be absent but may be categorized otherwise (e.g., as friendship). In addition, through learning, some stimuli (e.g., sexual) may develop less incentive value than other stimuli (e.g., cake).

Importantly, I do not offer this model as a perfect, universally applicable one. Rather, I believe it offers one way of thinking about (a)sexual development that is consistent with the scientific research (read critically) and less likely to produce views of asexuality and other nonsexualities as pretense, deficiency, or dysfunction. If this model is acceptable to people in a variety of disciplines, including science and medicine, it could help to encourage a more positive view of asexuality in these fields, perhaps tempering current biases. However, as has been demonstrated throughout this book, academics and activists must be cautious in basing positive political projects on scientific evidence. As was discussed in chapter 2, many psychologists and neuroscientists claim to use an incentive motivation model of sexual desire but still pathologize nonsexuality. Again, I see the greatest value of scientific research to ace political projects as its ability to clear conceptual space for thinking otherwise.

Feminist and Queer Strategies for an Ace-Affirming Future

If rereading scientific research on sexuality allows us to "denaturalize" the categories of sexual activity and desire, what implications does this have for intersectional feminist and queer ace projects? Here I discuss two starting points for an ace political project within feminist and queer studies, as well as how they might be bolstered by disrupting the categories of sexual activity and desire.

One starting point is the long-standing acknowledgment of and respect for benign sexual diversity within queer studies, explicitly articulated by Gaye S. Rubin in her field-founding essay, "Thinking Sex: Notes for a Radical Theory of the Politics of Sexuality" (1984). That this recognition of sexual diversity potentially includes a recognition of asexuality and nonsexuality is most explicitly stated by Eve Kosofsky Sedgwick in her introduction to *Epistemology of the Closet* (1990), in which she offers a series of "axioms" for queer theory. The first is deceptively simple: People are different from each other. Sedgwick goes on to provide a list of just some of the ways in which people may differ in the area of sexuality, which includes thirteen bullet points, for example: "Sexuality makes up a large share of the self-perceived identity of some people, a small share of others'"; "Some people spend a lot of time thinking about sex, others little"; and "Some people like to have a lot of sex, some little or none" (22–25). This starting point is generative for ace projects, as it allows for levels of attraction to be merely one of the ways in which people differ from each other in the area of sexuality, while challenging the hierarchical valuation of different forms.

While this is a productive project, there are also some limitations to locating feminist and queer ace projects within the opening provided by Rubin and Sedgwick. These come from understanding asexuality as simply another form of queer sexuality—as do many in contemporary aspec communities—rather than as a challenge to the coherence of sexuality itself as a category. Queer theory and sexuality studies are caught in the same necessarily paradoxical situation as other academic projects—like women's and gender studies—that seek to document the material effects of social categories. In documenting the effects of sexuality as a social category, sexuality studies simultaneously challenges sexual oppression *and* consolidates the importance of sexuality in society. The very call for sexuality studies as an academic field, as well as the proliferation of scholarly work on the topic, serves, both intentionally and unintentionally, to secure sexuality as a socially significant category—one that is, itself, a patriarchal, colonial, and racist invention. In part, it is the weight given to sexuality by US society that makes it an oppressive category in the first place.[8] There is, as far as I can see, no productive way to avoid this paradox—I certainly have not done so in this book. Academics and activists can only hope that our efforts to dismantle sexual oppression are at least slightly more effective than the (often unintentional) consolidating work that is a material result of our efforts.

An antinaturalist approach to sexual activity and desire encourages us to keep this paradox in mind. In an antinaturalist intersectional ace political project, three claims would be advanced simultaneously: (1) people are different in terms of their interest in sex (from Sedgwick); (2) differences in "sexuality" (including levels of sexual interest) matter, because sexuality is a vector of oppression (also from Sedgwick); and (3) we could be otherwise, such that sexuality did not matter more than any other difference, either at an individual or a societal level (the contribution of an antinaturalist approach). An antinaturalist approach to intersectional ace political organizing holds on to the possibility of creating a world in which whether one is interested in sex or not does not make a much of a difference unless a person wants it to.

A second queer feminist starting point for an intersectional ace political project is Audre Lorde's concept of the erotic, which has been a powerful resource for queer feminist scholars theorizing asexuality (Przybylo 2019; Owen 2022; Snaza 2020). In her essay "The Uses of the Erotic: The Erotic as Power" (1984), Lorde offers the erotic as an answer to a question that has troubled feminist theorists for decades: Under conditions of domination, how do members of oppressed groups become aware of their oppression, and how do they envision change? According to her, after experiencing the erotic, "we begin to demand from ourselves and from our life-pursuits that they feel in accordance with that joy which we know ourselves to be capable of. Our erotic knowledge empowers us, becomes a lens through which we scrutinize all aspects of our existence" (57). In other words, the erotic survives under oppression and can be experienced by members of oppressed groups. Once experienced, we can realize that systems of oppression are preventing us from fully experiencing the joy that we are capable of, and we can work to change the world in ways that will allow us to experience it more fully.

Yet, what does Lorde mean by "the erotic"? Her definition is in many ways ambiguous: "The very word *erotic* comes from the Greek word *eros*, the personification of love in all its aspects—born of Chaos, and personifying creative power and harmony. When I speak of the erotic, then, I speak of it as an assertion of the lifeforce of women; of that creative energy empowered, the knowledge and use of which we are now reclaiming in our language, our history, our dancing, our loving, our work, our lives" (55). Here Lorde suggests that she is not necessarily using the word *erotic* to describe a specifically sexual energy but rather a "lifeforce" or a "creative energy." Later she defines "the erotic" as an internal feeling of joy experienced when engaging in specific activities or through sharing a deep connection (whether physical, emotional, or intellectual) with another person: "There is a difference between painting a back fence and writing a poem, but only one of quantity. And there is, for me, no difference between writing a good poem and moving into sunlight against the body of a woman I love" (58).

Not surprisingly, this concept of the erotic—in which the same energizing joy can be produced by a variety of activities, including those traditionally thought of as sexual or nonsexual—has served as a significant resource for feminist and queer theorists thinking about asexuality. For instance, Ela Przybylo (2019) draws on Lorde's concept of the erotic to think about asexual erotics. According to her, Lorde's work, "opens up space for a deep intimacy that is not reliant only on sex and sexuality for meaning but that finds satisfaction in a myriad of other activities and relationships to the self and to others" (22). Ianna Hawkins Owen (2022) also turns to Lorde's discussion of the erotic to theorize about asexual and sexual pleasures: "In claiming the difference [between sex and painting a fence, for example] is one of quantity, not quality, Lorde might free us, however briefly, from an interpretation of a hierarchy of pleasures, while also holding space for mundane joy of which there is no subordination to other forms of joy—just more or less of them" (103). For both Przybylo and Owen, then, Lorde's concept of the erotic provides a way to value "nonsexual" forms of relationality and pleasure as equal to "sexual" forms and thus allows for an understanding of asexuality not as lack but as an attention to different but equally valuable forms of relationality and pleasure.

I too find Lorde's work enormously productive for theorizing about asexualities. Her assertion that multiple forms of pleasure and types of relationships can produce joy and energy is perhaps the most generous and helpful prototheorization of asexuality offered prior to the rise of contemporary asexual identity movements and associated scholarship. As I argued in chapter 2, some neuroimaging research on sexuality seems to reflect a fear of exactly the kind of "undifferentiated pleasures" that Lorde embraces in her discussion of the erotic. As I also suggested, Lorde's discussion of the erotic accords well with other modern neuroscientific studies on pleasure, which has found that different types of activities, such as eating, sexual activity, and drug taking, activate a common reward system.

However, an antinaturalist approach to sexual desire and activity suggests some limitations of Lorde's approach for an intersectional ace political project. Commentators have already debated whether her identification of the erotic as a specifically *female* energy entangles her work in gender essentialism (Ginzberg 1992; Carr 1993; Morris 2002; C. Ward 2023). There is also a sense in which Lorde's discussion of the erotic seems to set it up as some kind of essential internal drive ("lifeforce"), bearing some similarities to Freud's notion of the life drive, which he also called "Eros" (and opposed to the death drive), although this is also open to debate (for a discussion of Freud's life drive, see Kli 2018). Although the erotic in Lorde is not limited to the sexual, an antinaturalist intersectional ace project is suspicious of universal drive models in general.

Another limitation of an "eros approach" to ace activism and theorizing is tied to the concept of undifferentiated pleasures. On the one hand, this concept is highly appealing for ace political projects, as discussed above, and seems to accord in many ways with the arguments I have advanced thus far in this book. It acknowledges that there are not necessarily essential differences between the pleasures experienced as sexual and those experienced as nonsexual and works to disrupt the sexual/nonsexual binary. The concept also challenges the privileging of sexual over nonsexual pleasures. However, the equality of pleasures offered by Lorde's notion of the erotic is an equality dependent on sameness, that is, writing a poem and having sex are qualitatively the same, and this sameness makes them equal. If Rubin and Sedgwick recognize sexuality and other pleasures as different and unequal (sexuality is made less than by what Rubin calls a sex-negative society), Lorde offers a vision of them as the same and equal. Basing a political project on sameness (e.g., men and women are the same and therefore equal) has long been criticized by feminist theorists, most famously by Joan Scott in her article "Deconstructing Equality-Versus-Difference: Or, the Uses of Poststructuralist Theory for Feminism" (1988), in which she offers a way to move beyond difference/hierarchy vs. equality/sameness: "The critical feminist position must always involve *two* moves. The first is the systematic criticism of the operations of categorical difference, the exposure of the kinds of exclusions and inclusions—the hierarchies—it constructs, and a refusal of their ultimate 'truth.' A refusal, however, not in the name of an equality that implies sameness or identity, but rather (and this is the second move) in the name of an equality that rests on differences—differences that confound, disrupt, and render ambiguous the meaning of any fixed binary opposition" (48). In this chapter, I offered a model of sexual desire combining elements from the incentive motivation model of sexual desire and developmental systems theory, which provided an explanation for how some sensations, pleasures, activities, and relationships might come to be experienced as "sexual" (or not) and might come to hold motivation value (or not) over time through ongoing interactions with others and with social norms. It is my hope that this model offers a nonhierarchical way to think about different types of relationships and pleasures and how they might come to have different subjective meanings for different people, such that some activities may be experienced as sexual by some but not others and some may be motivating for some but not others. It is my hope that this model offers a way to equally value a variety of relationships and pleasures without needing to assert that they are the "same."

An additional limitation of Lorde's understanding of the erotic for ace political projects is closely related to the previous point. Not only does her notion of the erotic unite different types of activities, pleasures, and relationships under a unitary category ("sameness"), her choice of the term *erotic* indicates that this is a distinctively sexual category. Although Lorde claims that the term is not distinctively sexual, in fact it had strong sexual connotations in ancient Greece and still does in contemporary US society.[9] While uniting "sexual" and "nonsexual" pleasures in one category both destabilizes the sex/not-sex binary and in some ways "desexualizes" the activities, relationships, and pleasures that we traditionally think of as sexual, it also continues to expand the domain of the sexual. In a sense, uniting sexual and nonsexual phenomena using a term with sexual connotations suggests that the "nonsexual" can only be equal to the "sexual" if it becomes sexual. Thus, while Lorde's notion of the erotic has served as an important starting point for the development of ace political projects, I argue these should be placed alongside one that is more explicitly antinaturalist.

Toward an Antinaturalist Intersectional Ace Political Project

Using scientific research on sexuality to disrupt the sexual/nonsexual binary offers space for the development of a third possibility for an intersectional ace political project, adding to but not replacing those outlined above. Queer and sex-positive feminists have potentially destabilized the category of the "sexual" by expanding it, seeking recognition for a variety of sexual activities, not just sex for reproduction. The concept of "asexual erotics" has potentially destabilized the category of the "sexual" by combining it with the category of the "nonsexual," maintaining that all energizing pleasures and life-giving relationships are qualitatively the same. An intersectional feminist queer ace political project adds a third prong of attack on the colonial category of the "sexual": rejecting the "sexual" and "nonsexual" as relevant overarching categories of classification, instead developing more nuanced ways of thinking about a diversity of pleasures, activities, and relationships. This would recognize that, for example, an activity may provoke more than one feeling and serve more than one goal at a time, that the "same" activity may provoke different feelings and serve different goals for different people (or for the same person in different contexts), and that the value of an activity should not be determined based on whether it is sexual or not but by some evaluation of its personal, interpersonal, and societal effects.

We need a more nuanced and nonhierarchical way of thinking about the diversity of pleasures, bodily activities, bodily interactions, and emotional and cognitive relationships that make up our lives. Sex—alone or with others—is merely one of many activities that can allow (some of) us to experience pleasure, relaxation, and/or connection, but also pain, anxiety, and/or isolation. Other activities that can do the same include dancing, exercising, drug taking, cuddling, eating, laughing, sunbathing, solving a puzzle, giving or receiving a massage, and meditating. Ultimately, what really matters is not whether these are sexual or nonsexual or the same or different but whether we are satisfied with the quantity, quality, and diversity of experiences of pleasure, relaxation, and connection in our lives.

How would this translate into practice? Imagine the context of a therapy session. Rather than asking, for instance, "Are you satisfied with your sex life?," instead the therapist might ask, "Are you satisfied with the quantity, quality, and diversity of pleasurable experiences in your life?" Imagine the context of the *Diagnostic and Statistical Manual of Mental Disorders*: It would make no sense to have a category like "sexual interest/arousal disorder" in isolation any more than it would to have a category like "exercise interest/ arousal disorder" (for people who don't desire exercise and/or don't feel excited or aroused by the prospect of exercise). At most, if we remain committed to diagnostic categories (itself doubtful), we might craft some kind of diagnosis for people who are not able to derive pleasure from their lives and are dissatisfied by its absence and/or who once experienced pleasure from a diversity of activities but no longer find any pleasurable and are dissatisfied with the change.

Just as "sexual" or "nonsexual" would no longer be a relevant way of characterizing activities under this alternative way of thinking, these terms would no longer be a relevant way of categorizing relationships. There are so many different types that can enrich and/or impoverish our lives, and they can meet (or fail to meet) so many different emotional, intellectual, physical, and economic needs. Rather than asking, "Is this relationship sexual or nonsexual?," under this alternative way of thinking, we would ask, "What is this relationship bringing into my life, and what am I bringing into the life of this person I am in relationship with? And, looking across all of the relationships in my life, in total are my individual needs for partnership, support, communication, sex, emotional and physical intimacy, caregiving and care-receiving, validation, fun, pleasure, challenge, safety, understanding, learning, growth, and more being met by my relationships with others?"

Again, how might this translate in practice? As Sedgwick (1993) contends, in US society, most of our needs for physical and emotional intimacy are supposed to be met through one long-term, heterosexual, reproductive, monogamous, sexual, and romantic relationship, the type that is socially and legally recognized through the institution of marriage. Although the hegemony of marriage has been challenged by queer and feminist activists of various stripes (polyamorous, single by choice, etc.), it persists in contemporary US society. Aspec individuals and communities have developed their own challenge, for example, through the development of the term *queerplatonic*, which is often defined as a nonsexual and nonromantic relationship that involves the level of emotional intimacy, commitment, and prioritization typically associated with a long-term committed romantic and sexual relationship ("Queerplatonic Relationship," n.d.). In some cases, the term is used less to describe a specific type of relationship and more to describe one that doesn't fit into traditional categories. For instance, one aspec website says they are "relationships that purposely defy relationship categories, and can mix elements from platonic, romantic, and sexual relationships. They are each unique depending on the people involved in them, but they often involve some level of commitment or intimacy" (Ace and Aro Advocacy Project 2022). However, the use of the word *platonic* as part of the term still indicates that the concept is as much defined by what it is not (sexual and/or romantic) as what it is, and the (non)sexual/(non)romantic is still fundamental to its definition. Rather than seeking separate recognition for important but nonsexual and nonromantic relationships, an antinaturalist ace project suggests that important relationships should be recognized and supported by society (whether they are sexual, romantic, and/or otherwise, with human or nonhuman others).

Again, this way of thinking, which I have called an antinaturalist, intersectional ace political project, is not intended to stand on its own. Rather, it can work in conjunction with queer feminist efforts to challenge sexual oppression and obtain recognition for a diversity of sexual activities, relationships, and desires. It can also work in conjunction with a notion of "asexual erotics," which revalues both the sexual and the nonsexual by rendering them qualitatively equivalent. It is a project that may find affinities with others focused on relationality in ways not dependent on settler sexuality, such as Indigenous feminist projects of "making kin" (TallBear 2022), queer relationality (Bradway and Freeman 2022), multispecies kinship (M. Han 2022), and women of color othermothering (Turner 2024), to name a few. The ultimate goal of this intersectional, queer, feminist ace political project is to foster a society in which one's interest in sex and engagement in sex matters only to the extent that one wants them to matter. This kind of society would be more ace friendly, and simultaneously, allow for a greater diversity of human (and nonhuman) flourishing.

NOTES

Introduction

1. I use the term *hegemonic* as it is used in masculinity studies. Connell (2005) argues that there are multiple masculinities, which are inherently relational to each other and to multiple femininities. Drawing on Gramsci's discussion of hegemony, Connell describes "hegemonic masculinity" as the "configuration of gender practice which embodies the currently accepted answer to the problem of the legitimacy of patriarchy, which guarantees (or is taken to guarantee) the dominant position of men and the subordination of women" (77). I find this helpful in thinking about "science" as a particular form of knowledge production that exists in a relation to other forms and secures the dominant position of a particular conception of the world emerging from "Enlightenment" Europe (e.g., focused on rationalism, Cartesian dualism, and individualism), tied closely to intersecting systems of colonialism, anti-Black racism, patriarchy, heterosexism, and ableism.

2. Describing scientific findings as distinct from the material world they seek to capture is an oversimplification. As feminist science studies scholar Karan Barad (2007) argues, the material world (which she calls "phenomena") and the instruments we employ to apprehend it (which she calls "apparatuses") are both material-discursive and always "intra-acting"—a neologism she coined to describe the way phenomena and apparatuses do not preexist their interaction but are always already coconstitutive in ongoing, iterative interactions.

3. Of course, it is likely that asexual or nonsexual experiences, including a lack of sexual desire or activity, have existed throughout human history in various forms. However, an explicit articulation of asexuality as a sexual identity with a common definition, in addition to a community of people who identify as asexual, is a relatively recent phenomenon.

4. For example, according to the 2008 AVEN community survey, 46.6 percent of respondents were from the United States, 24 percent were from the United Kingdom, and 11.1 percent were from Canada (AVEN, n.d.-a). According to the 2020 community survey, 47 percent of respondents were still from the United States, but the percent from the UK (9.2 percent) and Canada (6.2 percent) had decreased, and that from other countries, including Argentina, Mexico, and Poland, had increased. The AVEN community survey is in English, so it under-represents non-English-speaking people who identify as asexual, although for the 2020 survey, four translation guides (German, Italian, Spanish, and

Portuguese) were offered (I believe the survey itself was only offered in English, but the explanation about this is a bit unclear) (Hermann et al. 2022). As of June 2023, AVEN had functional links to sites in the following languages: German, Polish, Dutch, Spanish, Italian, Finnish, Czech, Turkish, Russian, Chinese (2), Portuguese, and Danish, as well as to an India-specific site in English.

5. For example, asexual activist Yasmin Benoit was the grand marshal of the NYC Pride Parade in 2023 (Baska 2023). AVEN has a forum for organizing local in-person meetups.

6. In some instances, AVEN defines asexuality as a sexual orientation, while in others it defines it as an identity label. The home page of AVEN states, "Unlike celibacy, which is a choice to abstain from sexual activity, asexuality is an intrinsic part of who we are, just like other sexual orientations." However, under "About Asexuality," the site states, "Asexuality is like any other identity— at its core it's just a word that people use to help figure themselves out, then communicate that part of themselves to others. If you find the word *asexual* useful to describe yourself, you may certainly identify as asexual."

7. Asexual communities utilize the term *sexual attraction* to refer to what scientists call "sexual desire," and the terms *libido* and *sexual desire* to refer to what scientists call "sexual arousal." For instance, the AVEN website defines sexual attraction as the "desire to have sexual contact with someone else or to share our sexuality with them." For different histories of the development of AVEN's terminology, see Hinderliter 2009; Siggy 2022.

8. The differentiation between sexual and romantic attraction is sometimes called the "split attraction model" or the "differentiated attraction model" in ace and aro spectrum (aspec) discourse. For an ace history of the "split attraction model," see historicallyace 2016.

9. In general, sociological and psychological studies have concluded that asexuality-as-identity has much in common with other minority sexual identities or orientations, particularly in terms of identity development (e.g., MacNeela and Murphy 2015; S. Scott and Dawson 2015; Mollet 2020; Dawson et al. 2016; Winer et al. 2022; Copulsky and Hammack 2023; Carrigan 2011; Scherrer 2008; Robbins et al. 2015; Foster et al. 2019; Cuthbert 2015; Mitchell and Hunnicutt 2019; Vares 2018; Winter-Gray and Hayfield 2021; Kelleher and Murphy 2022; Houdenhove et al. 2015; Gupta 2017).

10. Scholarship examining asexual identity outside of the US, Canada, and the UK includes D. Wong 2014; Zheng and Su 2018; Kurowicka and Przybylo 2020; Batričević and Cvetić 2016; D. Wong and Guo 2020; Dana 2020.

11. According to the 2020 Ace Community Survey, 42.2 percent of respondents identified as gender nonbinary in some way and 15.2 percent identified as transgender (gender identity and transgender identity were separate questions) (Hermann et al. 2022). According to the 2015 National Transgender Discrimination Survey, 10 percent of trans respondents identified as asexual (Sandy James et al. 2016). In a recent study, 5.4 percent of transgender respondents identified

as on the asexual spectrum compared to 0.6 percent of cisgender respondents (Reisner et al. 2023).

12. Simultaneously, people with disabilities, especially women, have experienced high levels of sexual violence (Benedet and Grant 2014; Malihi et al. 2021).

13. Black feminist scholars have explored the ways in which Black women have been both hypersexualized and asexualized in white supremacist societies. On the one hand, since the beginning of the slave trade, European and white settler societies have portrayed Black women as hypersexual (Hammonds 1994). At the same time, Black women have in some cases been stripped of their gender (e.g., Hortense Spillers suggests that the Middle Passage involved a process of ungendering through which Black bodies were made flesh), and Black women were also portrayed as asexual "mammies" (Spillers 1987; Wallace-Sanders 2008). Black women, including Black feminist activists and academics, have responded to these conflicting stereotypes in a variety of ways (see note 14 below). For a discussion of stereotypes of Asian and Asian American men's sexuality, see Eng 2001; Le Espiritu 2004; Shek 2007; M. Lee 2020.

14. Black women, including Black feminist activists and scholars, have responded to conflicting stereotypes about Black women's sexuality in a variety of ways. In the early twentieth century, Black women reformers promoted public silence on sexuality in order to combat hypersexualized stereotypes and to protect their inner lives (called the "politics of silence," a "culture of dissemblance," and "respectability politics" by Black feminist scholars) (D. Hine 1989; Higginbotham 1992, 1994). This public silence around Black women's sexuality continued for decades (Hammonds 1994). Since the 1990s, a number of Black feminist activists and academics have articulated a positive discourse about Black women's sexuality and pleasure (e.g., see Morgan 2015; Nash 2014; Horton-Stallings 2007). The hypersexualization and asexualization of Black women, as well as their responses to this history, have made it difficult for them to claim a positive asexual identity. Ianna Hawkins Owen (2018) posits that the figure of the mammy may be read as intentionally "saying nothing" and thus brings into focus a Black asexual agency that Owen calls a "declarative silence."

15. As TallBear utilizes the term *compulsory settler sexuality* to describe a much broader set of norms and institutions beyond sexual disinterest, I continue to use the term *compulsory sexuality* in this book, while acknowledging that the *sexuality* part is coconstituted with patriarchal, racist, colonialist, ableist, heterosexist, and classist norms and institutions.

16. For discussions of sovereign erotics, see Rifkin 2012. For a brief discussion of the connections between asexualities and sovereign erotics, see the introduction to Przybylo 2019. For a discussion of "the erotic of abstinence," see Houar 2022.

17. For examples of analyses of compulsory sexuality, see, for media, Barounis 2014 and Boislard et al. 2022; for law, Emens 2013; for education, Mollet and Lackman 2019; and for religion, Brandley and Spencer 2023.

18. For an early critique of scientific research on asexuality, see Przybylo 2012.

19. The distinction between human and nonhuman has always been a political one. As a number of scholars have asserted, people of color, people with disabilities, and other minoritized groups have long been denied inclusion in the category of the "human" by dominant groups, ideologies, and institutions (McKittrick 2015; Parker 2018; M. Chen 2012). A number of European scientists in the eighteenth and nineteenth centuries claimed that Africans and other people of color were a separate (and inferior) species (J. Jackson and Weidman 2004; Dennis 1995; Schiebinger 1993). People of color and people with disabilities are still often viewed as "less human" than white, able-bodied people today (M. Chen 2012). In addition, a number of thinkers have argued that the human/nonhuman distinction reinforces speciesism (Horta and Albersmeier 2020; Wyckoff 2014; Westerlaken 2020). In this book, I use the terms *human animals* and *nonhuman animals* to indicate similarities and differences between those categories. I do not intend to imply a hierarchical relationship between them; however, I recognize that pairing these terms also suggests a binary opposition, simultaneously erasing commonalities and differences within the categories. My use of them is not intended to indicate that nonhuman animals do not have agency or do not deserve rights. I support scholarly efforts to distinguish between "the human" as a species category and "personhood" as a category that could be applied to nonhuman animals as well as other living and nonliving entities, while also recognizing the limitations of this approach (Korsgaard 2013; Wallis 2009; Riddle 2014; RiverOfLife et al. 2021).

1. "There Is a Great Deal of Denial in This Population"

1. While I focus on sexual disinterest, many other forms of nonsexuality were also pathologized during this period, including celibacy, sexual incapacity (i.e., impotence or erectile dysfunction), and nonenjoyment of sexual intercourse (i.e., orgasmic disorder). In some instances, one diagnosis included multiple types of nonsexuality: for example, the category of "frigidity" sometimes described disinterest in sexual activity and also nonenjoyment of sexual activity. In these instances, I examine the diagnostic category that encompasses sexual disinterest. In other instances, sexual disinterest was separated from other forms of nonsexuality, in which case I focus on the diagnostic category specific to sexual disinterest. My focus is on European and American medical professionals, as their hegemony means that their work has had a disproportionate influence on people living in many parts of the world. In addition, I focus only on key moments and key people, rather than attempting a comprehensive history. I also focus only on expert or prescriptive literature, which can help us to understand how sexual norms have been promoted by certain elite segments of society but generally does not provide much information about how patients interpreted these diagnoses or how they affected the lives of ordinary

people (Neuhaus 2000). A number of scholars have examined "nonelite" texts to learn more about how nonelites interpreted sexological ideas. For instance, Peter Cryle (2008) examines the concept of frigidity in a series of "middle-brow novels" published in Paris in the late nineteenth century. Aaron J. Stone (2023) argues that African American novelists in the early twentieth century challenged the racism of American sexology by using literature to present their own studies of sexual behavior.

2. Of course, any effort to repurpose a diagnostic category as a liberatory identity is fraught; Foucault famously makes this point in *The History of Sexuality* about the category of the "homosexual" (1990, 101).

3. From the start, sexology was a transnational and colonial project. In their edited volume on sexology as a global project, Fuechtner, Haynes, and Jones make three claims: "(1) that European sexual science was constituted on the basis of conceptions of Others considered outside of 'modernity'; (2) that actors outside of Europe became important interlocutors in a globalizing sexual science through 'unruly appropriations' of the field's emergent ideas; and (3) that ideas of sexual science circulated multidirectionally through intellectual exchange, travel, and internationally produced and disseminated publications" (Fuechtner et al. 2018, 3; see also Bauer 2015; Kahan and LaFleur 2023).

4. For another discussion of the history of the pathologization of sexual disinterest, see Kim 2014.

5. My approach is somewhat different than that of Peter Cryle and Alison Moore in their book *Frigidity: An Intellectual History* (2012). Cryle and Moore in some instances trace the history of a word (*frigidity*) and its shifting definitions and in some cases trace the history of a "referent" ("non-receptivity of a person to a particular desire or sexual expectation of another") and its shifting linguistic signs. In this chapter, I focus on tracing the history of a referent (disinterest in sex), but in doing so I am not suggesting that this is a stable phenomenon.

6. For a discussion of stereotypes of Asian and Asian American men's sexuality, see Eng 2001; Le Espiritu 2004; Shek 2007; M. Lee 2020. For a discussion of stereotypes of disabled sexuality, see Esmail et al. 2010; Shakespeare 1996; Tepper 2000. For a discussion of stereotypes of the sexuality of older people, see Walz 2002; Deacon et al. 1995. It is an oversimplification to say that these groups have only been asexualized; for example, people with disabilities have been portrayed both as asexual innocents and as sexual predators (Campbell 2017).

7. Black women provide an example of a group that was simultaneously hypersexualized and asexualized in white supremacist societies. For instance, as argued by Evelyn Hammonds (1994), in the nineteenth century, as a result of scientific racism, Black women's bodies were presented as sexual objects, and Black women as a group were thought to possess primitive sexual appetites. At the same time, Black women have in some cases been stripped of their gender (e.g., Hortense Spillers suggests that the Middle Passage involved a process of ungendering through which Black bodies were made flesh), and Black women

were also portrayed as asexual "mammies" (Spillers 1987; Wallace-Sanders 2008). As Black women as a group were both hypersexualized and asexualized, as noted above, they were unlikely to be individually diagnosed with a hypersexual desire disorder like nymphomania (Groneman 2001) or, as we will see in this chapter, a sexual disinterest disorder.

8. I agree with the historians who contend that there are ways to examine "asexual-like" experiences in the past that do not impose contemporary categories anachronistically, but this is not the focus of this chapter (see Bennett 2000 for a discussion of "lesbian-like"). How one defines asexual-like experiences—whether broadly as experiences of nonsexuality or narrowly as ones of disinterest in sex—will affect a history of asexual-like experiences. In either case, the history of asexual-like experiences will likely overlap with the history of celibacies. For histories of celibacy, see Abbott 2000; Kahan 2013; French 2021.

9. Aphrodisiacs were listed separately from cures for impotence (McLaren 2007).

10. By the end of his life, he agreed that homosexuality was not a disease, and he eventually supported the decriminalization of homosexuality (Oosterhuis 2000).

11. Krafft-Ebing does not specifically state that F. J. did not have sexual feelings toward the same sex, but presumably, if F. J. had experienced such feelings, Krafft-Ebing would have classified him as a case of "contrary sexual instinct."

12. A member of the Asexuality Visibility and Education Network translated into English some of the passages in Hirschfeld's work that "hint at asexuality" (tommy92 2014). I am grateful to Anson Koch-Rein for his assistance with translating passages from Hirschfeld's writing into English.

13. Hormone research at the beginning of the twentieth century also reflected concern with sexual disinterest, specifically in men. Starting in the 1880s, some early endocrinologists began to treat male sexual disorders through various operations, including injections of human and animal testicular material, transplantation of human and animal testicular tissue, and unilateral ligation of the vas deferens. These operations generally fell into disrepute by the 1930s. However, in that decade, testosterone was finally isolated, and doctors began to treat sexual disorders using synthetic testosterone administered by pills or injections. Proponents believed that these treatments could restore overall vitality, cure impotence, and improve libido (McLaren 2007). At the same time, doctors also began to treat women with hormones, particularly for symptoms of menopause (McLaren 2007). Thus, hormone replacement therapy (HRT) can be seen, to a degree, as part of the history of medical approaches to sexual disinterest.

14. For excellent histories of trans medicine, see Shuster 2021; Gill-Peterson 2018; Malatino 2020; Meyerowitz 2002.

15. Some modern translators have translated *automonosexualism* as "asexual" (Gherovici 2011, 5), but in my view, this is not accurate. The conjoining of *automonosexual* and *transvestite* by Hirschfeld has also been identified by Ray Blanchard as the precursor of his controversial concept of "autogynephilia,"

or the idea that some trans-feminine people experience sexual arousal at the idea of having a female body (Blanchard 1989, 2005). The concept of "autogynephilia" has been criticized by trans activists and academics, who believe that the category inappropriately sexualizes and pathologizes trans experience (Serano 2008, 2010, 2020).

16. Hirschfeld classified as heterosexual those who were assigned male at birth and were sexually attracted to women, and as homosexual those who were assigned male at birth and were sexually attracted to men, regardless of their gender identity (all of Hirschfeld's cases were assigned male at birth except one) (Hill 2005).

17. Different sexologists and sex therapists in the 1950s, 1960s, and 1970s classified the sexuality of trans people in various ways, ranging from three to five categories to a continuous scale. They also utilized the terms *asexual* and *automonosexual* differently, some using only one, some using them interchangeably, and some using them to mean two distinct things (for different examples, see Hamburger 1953; Randell 1959; Burchard 1965; Blanchard 1989; Benjamin 1966). When the diagnosis of "gender identity disorder" was added to the *Diagnostic and Statistical Manual of Mental Disorders* for the first time, in the *DSM-III* (1981), it included the following "subclassifications": asexual, homosexual, heterosexual, unspecified (Cohen-Kettenis and Pfäfflin 2009).

18. Although Krafft-Ebing believed in the existence of "true" or complete sexual anesthesia, many others, including Ellis, believed that there was no such thing. Rather, according to this position, almost all cases of anesthesia or frigidity are merely apparent—evidence of sexual interest would be found through closer examination, or the patient would reveal sexual interest with the right partner or stimulation (Ellis 1936). As we have seen, this is the position taken by Hirschfeld (1932). In general, this position allowed medical professionals to see almost all cases of anesthesia or frigidity as potentially curable.

19. It is interesting to note that the term *asexué* applied by the French officials is not the one used most often today by French asexual communities (who use *asexualité* and *asexuel[le]*). This demonstrates that simply tracing the historical use of a particular word would not reveal a complete history of asexual experiences. Thank you to Stephanie Koscak for assistance with translating the French into English.

20. However, these terms have been applied sporadically, up until the present. For example, in *The New Sex Therapy* (1974), Kaplan included a diagnostic category for women called "sexual anesthesia or conversions," which she defined as the failure to experience erotic feelings during sexual stimulation.

21. According to Laplanche and Pontalis (1973), Freud originally conceived of two groups of primal instincts: the ego or life instincts and the sex instincts. Later, Freud combined them into one group—the life instincts—and contrasted it with a second, the death instincts. At least in "Instincts and Their Vicissitudes" (1915), Freud leaves open plenty of room for ambiguity, contradiction, and contestation in his discussion of the primal instincts.

22. The most popular of these manuals included *Ideal Marriage: Its Physiology and Technique* (1926) by the Dutch gynecologist Theodore van de Velde and *Married Love* (1918) by the English paleobotanist Marie Stopes (McLaren 1999; Neuhaus 2000).

23. Many authors attributed female frigidity to both societal repression and the failings of men. Women were held partially responsible for overcoming their frigidity. However, most of the responsibility was given to husbands—according to the manuals, frigidity could be cured by teaching men to awaken their wives' sexual passions. Overall, these marriage manuals reflected concerns about changing gender roles in the interwar years. On the one hand, they acknowledged women's greater independence; on the other, they sought to shore up the institution of marriage and emphasized the role of husbands in managing their wives' sexuality (McLaren 1999; Neuhaus 2000).

24. The most well-known early advocates of this view include Wilhelm Stekel, Marie Bonaparte, and Helene Deutsch (Gerhard 2000; Moore 2009). These thinkers claimed that psychologically mature, feminine women should accept their passive role in sexual activity and give up the clitoris in exchange for the vagina as the center of their sexual life. Women who were unable or unwilling to do so were labeled "phallic" or "masculine" and diagnosed with frigidity. Treatments included psychoanalysis and, in some cases, surgical relocation of the clitoris closer to the vagina. This redefinition of frigidity in the 1920s and 1930s was in part a reaction to feminism, changing gender roles, and declining white birth rates, all of which were perceived as threatening to the established social hierarchy. White women who stepped outside of traditional gender roles, including feminists, lesbians, and those who lived independently from men, were labeled frigid by these thinkers (Moore 2009; Gerhard 2000; Jeffreys 1997).

25. The main architects of this definition in the United States were Edmund Bergler and Eduard Hitschmann, psychoanalysts and associates of Freud who immigrated to the United States from Austria; William Kroger, a gynecologist; and S. Charles Freed, an endocrinologist (Lewis 2010, 40–41; Gerhard 2000).

26. Irvine contends that Kinsey's notion of sexual capacity essentially served the same function as the concept of libido. According to Irvine (2005), Kinsey defined sexual capacity as something like an individual's "well-spring of sexual energy." People could have more or less sexual capacity, but, as noted below, Kinsey believed there is never a complete lack of sexual capacity (Kinsey et al. 1953, 374). In addition, while he critiques the use of the term *frigidity* for its connotation of "either an unwillingness or an incapacity to function sexually" (373), he still saw a lack of "sexual responsiveness" in women as a problem. According to him, women who appear to be frigid may simply need better sexual stimulation or "clinical help to overcome the psychologic blockages and considerable inhibitions which are the sources of their difficulties" (374). This indicates that Kinsey was not very far in his understanding of frigidity from many psychoanalysts.

27. The seven-point heterosexual-homosexual scale developed by Kinsey is used to classify a person on a continuum, from completely heterosexual (zero) to completely homosexual (six), based on their past sexual experiences and their "psychosexual reactions" to males and females. In addition, Kinsey also includes an "X" rating for those who "do not respond erotically to either heterosexual or homosexual stimuli, and do not have overt physical contacts with individuals of either sex in which there is evidence of any response" (Kinsey et al. 1953, 472). Although Kinsey found that very few men fit into this category, a significant percentage of women did: 14–19 percent of unmarried women, 1–3 percent of married women, and 5–8 percent of previously married women aged 20–35 (473–74).

28. In some instances, Kinsey is almost contemptuous of those who do not engage in sexual activity. In a section on single women, he notes that a significant proportion of them did not engage in any sexual activity leading to orgasm. Many were frustrated by their lack of a sexual outlet, but others were not. Kinsey claims that many of these women were teachers or academic administrators and/or were involved in women's clubs, service organizations, and efforts to shape public policy. Of these efforts, he writes: "When such frustrated or sexually unresponsive, unmarried females attempt to direct the behavior of other persons, they may do considerable damage. . . . If it were realized that something between a third and a half of the unmarried females over twenty years of age have never had a completed sexual experience, parents and particularly the males in the population might debate the wisdom of making such women responsible for the guidance of youth" (Kinsey et al. 1953, 527).

29. Their initial research subjects were sex workers; the later ones were primarily white, college-educated married couples (Tiefer 2004).

30. As suggested by a number of scholars, Masters and Johnston were interested in shoring up the institution of marriage by demonstrating the essential similarities between male and female sexuality (Irvine 2005; Tiefer 2004). Tiefer (2004) contends that their choice of subjects was potentially biased, as they only included women whose sexuality most closely resembled that of men (those who were able to orgasm regularly during masturbation and intercourse). Robinson (1976) argues that their biases also led Masters and Johnson to propose one model for male and female sexual response, despite the fact that they found a number of differences between male and female sexuality.

31. Some feminists in the 1970s (e.g., Koedt 1973) credited Masters and Johnson for challenging the concept of frigidity. Like Kinsey, they challenged the idea that women are sexually dysfunctional if they are unable to reach orgasm via vaginal intercourse alone, and they also did not view frigidity as the result of women refusing to accept feminine roles. However, they still saw the failure to reach orgasm after "appropriate stimulation" as a disorder, which they called "female orgasmic dysfunction" (Masters and Johnson 1970).

32. Second-wave feminists challenged the use of the term *frigidity* to pathologize women who did not experience orgasm as a result of vaginal stimulation during heterosexual intercourse, most famously, Anne Koedt (1973) in her article "The Myth of the Vaginal Orgasm." Due, in part, to feminist critiques, the term was abandoned by medical professionals, who adopted terms such as *orgasmic dysfunction* and *sexual arousal disorder* to describe an inability to experience sexual pleasure or orgasm, and defined these diagnoses in more gender-neutral ways (Elliott 1985). Interestingly, these more recent diagnoses are more similar to some of the earlier sexological categories than to the mid-twentieth-century psychoanalytic understandings of frigidity.

33. In her 1977 paper, Kaplan's citations only go back to Masters and Johnson. She does not reference any earlier conceptualizations of low libido as a disorder. She suggests that she and Harold Leif are identifying this disorder for the first time (Kaplan 1977). In *Disorders of Desire*, she claims that, up until a few years prior, sexual response was seen as a single event or clinical entity, and all sexual problems were lumped into one diagnostic category. In her reference section, she states that "the literature on disorders of sexual desire is rather scant" and lists only five sources, including her article, Harold Lief's, and the *DSM-III* (Kaplan 1979, esp. 3–7, 228). As we have seen, many sexologists clearly distinguished between desire disorders, arousal disorders, and orgasm disorders and identified low sexual desire as a disorder. This suggests that Kaplan did not read the research of sexologists. In his very short article on the topic, Lief mentions only Sandor Rado's concept of a "sexual motive state" as a precursor to his own ideas about sexual desire (Lief 1977).

34. According to Harold Lief, the change from "inhibited sexual desire" to "hypo-active sexual desire" was made in order to replace "psychodynamic" language with a more acceptable "phenomenological perspective" (Leiblum and Rosen 1988, viii). In *Disorders of Desire*, Kaplan (1979, 58) uses both "hypoactive sexual desire" and "inhibition of sexual desire (ISD)." She states that she utilizes the former when the cause of the desire disorder has not yet been identified and the latter when it has been established that sexual desire is inhibited by psychic factors.

35. The report is not public. I was given access to a redacted version by the authors.

36. Supporters of the gender-specific responsive desire model assert that it is feminist because it accepts as normal both "spontaneous" and "responsive" desire among women and only identifies them as pathological if they experience neither (previously they would be pathologized if they experienced responsive but not spontaneous desire) (Basson 2000, 2001; Brotto 2010b). Critics argue that the problem is that this model is applied to women only (men are still pathologized if they experience responsive but not spontaneous desire), as the gender-specific responsive desire model serves to reinforce the long-standing idea that women are sexually passive until their desire is stimulated by a male

partner. For critiques of this model, see Spurgas 2020, 2022. The responsive desire model is discussed further in chapter 3.

37. For the *DSM-5*, "gender identity disorder" was changed to "gender dysphoria," and the specifier for sexual attraction to male, females, both, or neither was removed. For a discussion, see Zucker 2015.

38. Homosexuality was listed as a mental disorder in the *DSM* until 1973, when it was removed by the APA. "Ego-dystonic homosexuality" was added to the *DSM* in 1980, a (controversial) diagnosis for those who were distressed by their same-sex desires. "Ego-dystonic homosexuality" was removed from the *DSM-III-R* in 1987 (Cabaj 2009; Margolin 2023a; Drescher 2015).

2. Sex in the Machine

Much of the thinking for this chapter was done in collaboration with Cyd Cipolla. As graduate students, Cyd and I received a grant from the Neuroethics Program at Emory University to teach a course titled "Feminism, Sexuality, and Neuroethics." We also coauthored an unpublished manuscript, "The Queer Feminist Neuroethics of Sexuality," some of which was published as the chapter "Neurogenderings and Neuroethics" in *The Routledge Handbook of Neuroethics* (Cipolla and Gupta 2017). I am grateful to Cyd for her friendship and intellectual collaboration, and for allowing me to use some of the unpublished manuscript in this chapter.

1. Komisaruk and Wise's first study involved mapping the clitoris, vagina, and cervix on the sensory cortex through fMRI (Komisaruk et al. 2011). Following this, they performed two studies, one investigating the activation of the sensory cortex by imagined genital stimulation, and one to compare brain activity at orgasm (self- and partner-induced) with that at the onset of genital stimulation, before the onset of orgasm, and after its cessation (Wise et al. 2017). I believe that the research from the third study was presented at the conference in 2011.

2. Sukel's posts were intended to serve as teasers for her 2013 book, *This Is Your Brain on Sex: The Science Behind the Search for Love*.

3. Komisaruk (2012) writes that he began his career studying rats, finding that vaginal stimulation might block pain response in female ones. After seeing his wife suffer pain from terminal cancer, he focused his research on the possible connection between vaginal stimulation and brain response in rats and humans. He obtained a patent for VIP, the peptide involved in the vaginal-stimulation-to-analgesia pathway in rats. From this, he moved to researching orgasm in women with spinal cord injuries and, from there, orgasm in "abled-bodied" men and women.

4. In one article, Komisaruk also mentions that he has utilized his research in his own sex life to "understand better how to elicit sexual pleasure in a woman," which raises all kinds of questions (Altman 2010).

5. This discussion of Komisaruk and Wise's research draws on a blog post I published in 2012 on *The Neuroethics Blog* (Gupta 2012a).

6. In this book, I do not offer an analysis of orgasm as a specific type of pleasure. For an analysis of the scientific debate about the evolutionary "purpose" of female orgasm, see Lloyd 2005. For other feminist and queer analyses of orgasm, see, for example, Bosley 2010; Willis et al. 2018; Gerhard 2000; Tuana 2004; Jagose 2012. For an asexuality studies discussion of orgasm, see Cerankowski 2021.

7. Specifically, critics have targeted multiple comparisons and the problem of circularity. According to Kriegeskorte et al. (2010), in neuroimaging, the problem is that testing across multiple brain locations (voxels) increases the chance that voxels will pass the significance threshold by chance even when there are no true effects ("false-alarm voxels"). However, researchers employ various statistical methods to address this problem. The problem of circularity arises in calculating "effect size" (or the magnitude of a finding), when scientists first select voxels of interest and then only calculate the effect size for selected ones, leading to selection bias.

8. This chapter employs the term *sex/gender*, as it is used by neurofeminist scholars to indicate that sex and gender are mutually constitutive (Jordan-Young and Rumiati 2012).

9. Rippon et al. (2014) identified the following specific limitations of contemporary sex/gender neuroimaging research: an essentialist view of sex/gender; routine comparisons of male and female subjects with only positive findings reported; biased institutional databases; small sample sizes; reporting all sex/gender differences without reporting the size of these differences; the use of gender stereotypes to interpret data; and the single "snapshot" approach to neuroimaging research, which identifies sex/gender differences at one point in time without explaining their origin or considering the possibility that they may change over time or context.

10. In order to measure sex/gender in a multidimensional way, Rippon et al. (2014) suggest that researchers could assess gendered experiences, socialization, behavior, and cognition rather than just asking participants to check male or female (Rippon et al. 2014). Persson and Pownall (2021) argue for the potential of open science (including preregistration, data sharing, and accountability) as a way to reduce sexism in neuroscience and psychology research.

11. For a discussion on the use and abuse of race and racialization in neuroscience research, see Kaiser Trujillo et al. 2022.

12. Since Dussauge and Kaiser's (2012) review, scientists have continued to produce neuroimaging studies examining various aspects of sexual orientation (e.g., Kagerer et al. 2011; Sylva et al. 2013).

13. For a critique of the diagnosis of hypersexuality, see Moser 2010. For a critique of the social and cultural construction of erectile dysfunction, see Potts 2000. For an overview of critiques of the category of pedophilia, see Malón 2012.

14. In the United States, the stereotype of the pedophile is of a white man (Chenier 2012; Livingston et al. 2024).

15. There are neuroscience studies that have employed other methods, for exam-

ple, eye-tracking, to explore sex/gender and racial/ethnic differences in sexual response, but the one study I examined in this category adopted the interpretation that women are essentially more sexually flexible than men and used broad generalizations in regards to ethnic categories, describing Western culture as sexually liberal and Asian culture as sexually conservative (Ganesan et al. 2020).

16. A few studies included older subjects. For instance, Martynova et al. (2017) used men aged 49–74 in order to "maximize exclusion of the effects of physiological arousal during viewing of erotic images" (394).

17. The dual control model of human sexuality does have a longer history. According to its authors, it is generally accepted that most brain functions involve both excitatory and inhibitory processes. They also identify a long history of research into the excitatory and inhibitory processes involved in sexuality in nonhuman animals (Bancroft et al. 2009).

18. One feminist concern about neuroimaging studies of sex, gender, and sexuality is that they do not investigate individual variability. A recent neuroimaging study by P. Chen et al. (2020) responds to this by assessing variability between subjects in response to watching erotic movies. While promising, the sample was a small number of male research subjects only. Another feminist critique of neuroimaging studies is that they tend to take a "snapshot" approach—imaging brain activation at one point in time without considering the possibility of change over time. One neuroimaging study of sexual response by Wehrum-Osinsky et al. (2014) seems to respond to this by imaging brain response to sexual stimuli at one point, and then again with the same participants 1–1.5 years later. The study should be commended for considering stability and change, for using a somewhat larger sample size of both men and women (56 subjects), and comparing response to sexual stimuli to response to nonsexual positive, negative, and neutral images. However, it also explicitly sets out to identify stability in neural response (as opposed to change) in order to contribute to a better understanding of sexual disorders. Thus, it contributes to larger efforts to distinguish between "normal" and "pathological" sexuality.

19. A few studies have measured brain response to smelling "human pheromones" (e.g., Berglund et al. 2006), while at least one has measured brain response to words spoken in "erotic prosody" (Ethofer et al. 2007).

20. Studies comparing brain response of heterosexual men to heterosexual women handle the issue of preferred sexual stimuli in different ways. Some show the male and female subjects different pictures (e.g., Hamann et al. 2004; Costa et al. 2003). Others show male and female subjects the same set of pictures or videos (e.g., Karama et al. 2002; Walter et al. 2008).

21. Exceptions include a study by Redouté et al. (2000), which compared brain activation in response to erotic videos, moderately arousing erotic pictures, and highly arousing erotic pictures. Another exception is one by Ferretti et al. (2005), which compared brain activation in response to erotic video clips of long duration to briefly presented erotic still images. A study by Bühler et al.

(2008) compared brain response to the same erotic stimuli presented in a block design vs. an event-related design. In block designs, stimuli are presented in blocks, with rest in between. In event-related designs, stimuli are presented as brief interleaved events. An interesting case is a study by Ortigue and Bianchi-Demicheli (2008), which compared brain activation in response to photos of people rated as highly desirable to those rated as mildly desirable. Interestingly, the authors describe the second set as "nondesirable stimuli" even though they had an average desirability rating of 4.28/10 (the first group had a rating of 7.52/10), using a scale of 1 (least desirable) to 10 (most desirable). The authors claim that their results demonstrate "two dissociable neural networks for [desirable stimuli] and [nondesirable stimuli]" (342). It is somewhat unclear how to interpret the results, however, if the stimuli are understood as highly desirable vs. mildly desirable (rather than highly desirable vs. not desirable). Studies comparing preferred vs. nonpreferred sexual stimuli are discussed elsewhere in the chapter.

22. Dussauge (2013) does not discuss the issue of preferred vs. nonpreferred sexual stimuli.

23. Gola et al. (2016) argue that the distinction is important in regard to the question of whether compulsive sexual behaviors share common brain mechanisms with other addictions. They note that in studies on substance and gambling addiction, cues evoke increased responses in the ventral striatum among addicted people, while rewards evoke stable or decreasing responses. Therefore, if researchers think VSS are cues, they would expect increased response in the ventral striatum among subjects with compulsive sexual behavior (CSB), whereas if they think VSS are rewards, they would expect stable or decreasing response in the ventral striatum.

24. According to a comprehensive review performed by Stoléru et al. (2012), most neuroimaging studies of sexual response up to that point had employed the subtractive method, in some cases supplemented by "the combination of different contrasts" and/or by correlational analysis between levels of brain activation and levels of a psychological or physiological measure of sexual arousal (1489). By a combination of different contrasts, they are referring to studies that, for example, compare brain sexual response between two or more groups of research participants. These studies first use the subtraction method for all groups—the comparison task is subtracted from the task of interest for all groups. Then the scientists compare the results of this subtraction between the different groups in the study. Stoléru et al. (2012) identify several other types or variations of studies performed: At least one employed a classical conditioning technique (Klucken et al. 2009), in which participants were shown one geometric figure before erotic stimuli and a different one before nonerotic stimuli. Some participants learned to associate the first figure with erotic stimuli, while others did not. In theory, this allowed researchers to compare brain activation to the same image, experienced as sexual by one group

and nonsexual by another. Several studies have utilized different methods to measure brain response to sexual stimuli over a particular time course (e.g., Ortigue and Bianchi-Demicheli 2008; van Lankveld and Smulders 2008; Ferretti et al. 2005; Feng et al. 2012). Several have employed different techniques to present sexual stimuli subliminally (e.g., Wernicke et al. 2017; Gillath and Canterberry 2012), discussed in regard to pedophilia. Stoléru et al. claim that, as of 2012, there was still a need for studies to investigate functional connectivity in regard to brain sexual response. In a review analyzing studies from 2012 to 2017, Ruesink and Georgiadis (2017) noted the continued dominance of activation over connectivity approaches (28 vs. 14 studies). Examples of studies that have used a connectivity approach include Seok et al. 2016; Klucken et al. 2016; S. Lee et al. 2015. A recent study by Klein et al. (2020) deemphasized the subtraction method in favor of connecting a person's brain activation patterns with their ratings of the vss in regard to valence and sexual arousal, and then correlated this association with their level of "problematic pornography use" (PPU).

25. One study by Seo et al. (2009) did include the following contrasts: activation in response to "erotic couple" pictures vs. "happy-faced couple" pictures, and activation in response to food pictures vs. nature pictures. However, the purpose of the study was not to compare the brain response to erotic images vs. food images; rather, it was to test the association between sexual hormone levels and brain response to erotic images. The food/nature contrast was included to demonstrate that sexual hormone levels were not associated with a different "drive" (hunger). In addition, the study was based on a small sample (12) of heterosexual men only.

26. The point is not necessarily to debate whether sexual pleasure is the same or different from other types of pleasure. Audre Lorde famously argued that seemingly different forms of pleasure—from writing a poem to painting a fence to lesbian sexual activity—are all "erotic" activity, differing not in quality but only in the amount of pleasure produced (Lorde 1984). For her, this conceptualization of commensurable pleasure is liberating, pushing back against commodified and patriarchal forms of sexuality. In her article, Dussauge (2013) explores the influence of neuroeconomics on neuroimaging research on sexuality, contending that neuroeconomics views all rewards as commensurable (able to be exchanged in a common marketplace, all possessing some level of value that can be calculated using common currency). For her, this conception is deeply problematic, consonant as it is with neoliberal capitalist values. Somewhat in contrast, my review of the neuroimaging research on sexuality in this chapter suggests that researchers generally conceptualize different rewards as producing both similar brain activation/deactivation patterns and unique activation patterns. Regardless, politically speaking, I argue that it is less important to adjudicate whether different forms of pleasure are essentially the same or different than to challenge social norms, practices, and institutions that create unjust

hierarchies between different forms of pleasure, a point I discuss further in the conclusion to the book.

27. This issue plays out somewhat differently for the neuroimaging studies that perform a "double" comparison (see note 24). For instance, when they compare brain sexual response in "healthy" women to that of "sexually dysfunctional" women, they generally first subtract the neutral task from the sexual task in both groups (first comparison) and then compare the remainder results for "healthy" women to those for "dysfunctional" women (second comparison). Any identified difference is a potential target for pharmaceutical intervention, and it does not necessarily matter if it is in a specifically "sexual" brain response system. However, when comparing brain sexual response in men to that of women, the scientists perform a similar double comparison (subtracting the neutral stimuli from the sexual stimuli and then comparing the results for men to the results for women) (e.g., Hamann et al. 2004). Researchers suggest that this isolates a difference between the brain sexual response of men and women. However, depending on what control task is used, the difference may not be specific to "sexual" response. For example, one eye-tracking study found that men looked more at images of women than of men, while women looked at images of men and women more equally, regardless of whether they were erotic (Lykins et al. 2008). Thus, if a neuroimaging study compares erotic images of people to control images without people, a sex/gender difference may be identified, but this might reflect a difference in viewing pictures of people rather than one in brain sexual response per se.

28. An unusual example is a study by Stark et al. (2005), which compared response to disgust-inducing images, "vanilla" erotic images, and sadomasochistic (SM) erotic images among subjects with and without SM interests. Not atypically, it found common patterns of activation for erotic and disgust responses, as well as unique areas of activation for each.

29. A study by Heinzel et al. (2006) did measure brain response to erotic-emotional, nonerotic-emotional, and neutral (erotic and nonerotic) images. However, it included only a small sample (13) of male-only participants. In addition, its purpose was to isolate the neural correlates of "self-relatedness" in response to erotic-emotional images from those in response to nonerotic-emotional ones. A study by Sabatinelli et al. (2007) did compare response to pleasant images (which included erotic images and those of romantic couples) to neutral ones (neutral people or dental images) and unpleasant ones (mutilations, snakes, or threatening people), but only examined the nucleus accumbens (NAc) and medial prefrontal cortex (mPFC).

30. There are at least two strange choices made here. One is the selection of sports images, as earlier in the paper, the authors noted that previous studies had used sports clips as control images but that these differed from sexual stimuli in general emotional dimensions, like induced feelings of pleasure. The other is the selection of images with and without bodies in the emotional/nonsexual

control group. Earlier, the authors noted that the brain responds specifically to human bodies and so argue for the importance of including them in the nonsexual emotional control stimulus. However, combining bodily and non-bodily images in the emotional group (while including only bodily images in the sexual group) seems to confound the procedure. Indeed, a confusing choice of language reflects this: the authors sometimes refer to the sexual images as "bodily images" and the emotional ones as "nonbodily images," even though the emotional images include pictures with and without bodies (Walter et al. 2008).

31. Scientists often select images from databases (e.g., the International Affective Picture System [IAPS]), which are usually rated for valence level (unpleasant to pleasant) and arousal level (calm to exciting). For a discussion, see Alarcão and Fonseca 2018. Because only the negative-context erotic and negative-context emotional pictures were rated similarly, the researchers are only able to make definitive claims about areas that were differentially activated by both positive-context erotic stimuli compared to positive-context emotional stimuli *and* negative-context erotic stimuli compared to negative-context emotional stimuli. They describe this as conjoining the contrast (positive erotic > positive emotional) with the contrast (negative erotic > negative emotional) (Walter et al. 2008).

32. Other studies that compare response to sexual images vs. emotional images include the following: Metzger et al. (2010) uses the same simulation paradigm as Walter et al. (2008) but with a higher-resolution fMRI scanner and a smaller group of male-only participants. Feng et al. (2012) employs EEG to measure brain response over time to erotic, nonerotic positive, nonerotic negative, and nonerotic neutral pictures. In this study, researchers sought to identify "implicit" brain response to the images by asking participants to judge the color of the frame after each viewing rather than asking them to respond to the content.

33. Like the other studies within the broader category of "difference studies," these are largely focused on either (1) identifying differences in brain activation patterns between subjects with different sexual orientations, and, in some cases, developing computer models capable of identifying a person's sexual orientation based on their brain activation in response to VSS (e.g., Ponseti et al. 2009); and/or (2) contributing to a broader research agenda focused on demonstrating that women are less "specific" in their sexual response compared to men (in other words, women, regardless of sexual identity or orientation, will supposedly respond sexually to people of any gender).

34. Theresa N. Kenney (2020) explores the overwhelming whiteness/Global North orientation of online asexual communities through their use of images of "Western" cake as well as the reorientation of the cake symbol by a Pilipinx artist who created images of *sapin-sapin* (Pilipinx rice cake) using "ace" colors.

35. Arguably, evolutionarily speaking, it was not necessary for all members of a species to connect genital stimulation to a pleasure/reward system. Both human and

nonhuman animals engage in sexual behavior for a variety of reasons, including pleasure, dominance, social bonding, and even, in some cases, protection or food (Clarke et al. 2022; Furuichi et al. 2014). Even if only the majority of a species experiences pleasure during sex, it seems possible that this would lead to sufficient reproduction. This allows for the possibility that "nonsexuality" (not experiencing either sexual pleasure and/or sexual desire) could persist among a minority of a species over generations without affecting its reproductive success.

36. As discussed in chapter 3, scientists claim that the sexual motivation of nonhuman animals can be assessed in the lab by allowing a test subject to choose whether to engage in sexual activity with "stimulus animals." Manipulating the testing environment and the test and stimulus animals has allowed researchers to explore the hormones, neurotransmitters, and genes involved in rat sexual motivation. According to leading scientists in the field, the evidence from these experiments provides support for an incentive motivation model of sexual desire (Ågmo and Laan 2022; Georgiadis et al. 2012). For rats, at least, there is evidence that many males and females without any sexual experience will prefer spending time with a sexually receptive rat of the opposite sex and/or with the odor of one, so at least some interest in "sexually receptive" odors is "unconditioned" (does not need to be learned through experience). But (pleasant) sexual experience generally serves to increase the motivation value of sexually receptive odors and sexually receptive rats of the opposite sex (Heijkoop et al. 2018; Ågmo 1999). But this is for rats. As Ågmo and Laan (2022) point out, there is no reason to believe that humans recognize "sexual incentives" as such without learning, whether direct or indirect.

3. Pandas, Voles, and Rams

Much of the early thinking for this chapter occurred when I was a graduate student at Emory University, earning a certificate in Mind, Brain, and Culture. Under the auspices of the program, I audited a course by and wrote a paper for Dr. Kim Wallen, some of which was used in this chapter. I am grateful to Dr. Wallen for generously allowing me to audit his course and for his willingness to engage with me about asexuality and "asexual phenomena" in nonhuman animals. I also want to thank Dr. Robert McCauley and Dr. Laura Namy, of the Emory Center for Mind, Brain, and Culture, for their enthusiastic support.

1. It is unclear whether breeders would actually show pandas videos of humans dressed as pandas having sex, as opposed to ones of actual pandas having sex.
2. The humorous intent behind the post is signaled by the text emoji XD.
3. The question of whether we can celebrate non-normative sexualities when they are the result of trauma, violence, and/or oppression is discussed later in the chapter.
4. In "How Meat Changed Sex: The Law of Interspecies Intimacy After Industrial Reproduction" (2017), Gabriel Rosenberg examines some of the politics

of artificial insemination practices in the meat industry. He contends that our society ostensibly forbids bestiality but actually allows for human/animal sexual interaction in the context of the meat industry, because it contributes to the reproduction of biocapital.

5. Not surprisingly, some media reports of this finding claimed that the missing ingredient in panda breeding in captivity had been "love" (Pappas 2015).

6. The focus of this chapter is sexually reproducing animals. Chapter 4 focuses on asexually reproducing organisms.

7. My use of the terms is not intended to indicate that nonhuman animals do not have agency or do not deserve rights. I support scholarly efforts to distinguish between "the human" as a species category and "personhood" as a category that could be applied to nonhuman animals as well as other living and nonliving entities, while also recognizing the limitations of this approach (Korsgaard 2013; Wallis 2009; Riddle 2014; RiverOfLife et al. 2021). It is beyond the scope of this chapter to explore what a truly ethical relationship between humans and nonhuman animals should look like; this question has been explored in depth by queer and feminist science studies scholars, who have developed concepts such as "multispecies justice" and "queer affiliations" to imagine more just relationships between human and nonhuman species (Weaver 2021; Chao et al. 2022)

8. Again, the question of whether we can celebrate nonnormative sexualities when they are the result of trauma, violence, and/or oppression are discussed later in the chapter.

9. For the purposes of this article, which seeks to establish stricter criteria for identifying homosexuality in nonhuman animals, Vasey (2002, 145) defines sexual behavior narrowly as "including courtship displays (or sexual solicitations), mounting, and any interaction involving genital contact between one individual and another. Although stimulation of the genitals or other erogenous zones can result in orgasm, orgasmic response is not a necessary criterion for labeling a behavior as sexual, nor is penetration." Vasey argues that some of the behavior that has been labeled "sexual" by other researchers may also simply be social behavior, so researchers must have some justification for why they are doing so.

10. There is evidence that some aspects of sexuality can be cultural in nonhuman animals; for example, sexual behavior varies in different groups of female Japanese macaques living in different areas, and these groups may socially transmit specific sexual patterns to their offspring (Leca et al. 2014).

11. A number of researchers maintain that tests like the sexual incentive motivation (SIM) test can identify the sexual motivations of nonhuman animals. For rats, during the SIM test, the rat is placed in an arena with small, attached cages. Inside each is a "stimulus" rat, including a "sexually receptive" rat of the "opposite" sex, a sexually nonreceptive rat of the "opposite" sex, and/or a rat of the same sex. Female rats are considered "sexually receptive" when they have

received estrogen and (usually) progesterone, which triggers lordosis behavior, which is a postural reflex thought to facilitate sexual activity (Tsukahara et al. 2014; Heijkoop et al. 2018). For male rats, they are generally considered sexually receptive if they are not castrated and sexually nonreceptive if they are, although (notably) the language of "sexually receptive" does not seem to be utilized; rather male rats are described as intact or "sexually active" (e.g., Snoeren and Ågmo 2014). To simplify, if the test subject rat spends more time near the sexually receptive rat of the "opposite" sex, this is thought to reflect the sexual motivation (or desire) of the former (Heijkoop et al. 2018; Ågmo and Laan 2022). Manipulating the testing environment, as well as the test and stimulus animals, has allowed researchers to explore the hormones, neurotransmitters, and genes involved in rat sexual motivation. According to leading researchers in the field, the evidence from these experiments provides support for an incentive motivation model of sexual desire, which was discussed in chapter 2 (Ågmo and Laan 2022; Georgiadis et al. 2012).

12. For examples of scientific studies on asexual phenomena in nonmammals, see Neal and Wade 2007; Lindzey and Crews 1992.

13. Neurons are thought to selectively express the gene c-fos when excited, which codes for fos protein products. Scientists remove the brains of nonhuman animals and then stain them with antibodies for those products. The number of "fos-immunoreactivity (IR)-positive neurons" (or the number that are stained by antibodies for fos protein products) is thought to reflect neural excitability in the brain region of interest (Iha et al. 2017).

14. The Asexuality and Visibility Education Network (AVEN) was founded in 2001 (Hinderliter 2009).

15. In my own research, I found that some interviewees who eventually went on to adopt an ace identity had originally considered medical explanations and/or treatments for their (a)sexuality (Gupta 2016b, 2017).

16. In my first book, *Medical Entanglements: Rethinking Feminist Debates About Healthcare*, I offer a more nuanced discussion of medical interventions. I argue that individuals, in consultation with their communities of choice, should be able to pursue medical interventions that they deem important for their own well-being, and that, rather than oppose specific interventions, feminist academics and activists should direct their efforts to changing the broader systemic factors that make most medical interventions so regressive, normalizing, and problematic in the first place (Gupta 2019b).

17. For a discussion of the use of nonhumans animals in scientific research on female hypoactive sexual desire disorder, see Potts 2022.

18. For a critique of sexual selection theory, see, for example, Conley et al. 2011. For a critique of brain organization theory, see Jordan-Young 2010.

19. For examples of scientific evidence for a biological basis of coercive sexual behavior in nonhuman animals, see Moldowan et al. 2020; Baniel et al. 2017. For an extended discussion of the evidence for a biological basis for coercive

sexual behavior in humans, see Thornhill and Palmer 2000. For a feminist discussion of this issue, see Vandermassen 2011.

20. Eve Sedgwick coined the terms *universalizing* and *minoritizing* to describe different approaches to sexuality. The universalizing approach understands homosexuality as a social/political category that is important in the lives of "people across the spectrum of sexuality." The minoritizing approach understands it as a distinct and stable sexual minority (Sedgwick 1990). Applied to asexuality, the latter approach would see it as only relevant to a small minority of people with asexual orientations, while the former would see it as relevant to everyone—first, because compulsory sexuality regulates everyone, and second, because asexuality's challenge to it could benefit many people beyond just those who identify as asexual.

21. In disability studies, scholars use the term *bodymind* to indicate the interdependence and inseparability of the body and mind (Schalk 2018).

22. Many scientists believe that some species of female primates can experience orgasm, but this, in itself, is an interesting area of research in need of close analysis. Initially, animal behaviorists observed that during reproductive copulation, female primates sometimes engaged in a pattern of reactions including "clutching," muscular body spasms, and characteristic vocalizations, which were identified as evidence of orgasm *because* they took place during copulation. Later researchers attempted to determine if this behavioral pattern included vaginal, anal, and uterine contractions; hyperventilation; and/or an increase in blood pressure and heart rate, which they maintain can be observed behaviorally. For instance, in a study of Japanese macaques, female orgasm was "defined as a clutching reaction associated with muscular body spasms and, sometimes, characteristic vocalizations" (Troisi and Carosi 1998, 1262). Scientists have also argued that female rats experience orgasm (Pfaus et al. 2016; see also Georgiadis et al. 2012).

4. Amoebas Are Us?

1. As will be argued later in the chapter, the scientific definition of "asexuality" is not consistent, and certainly includes organisms that reproduce (just not sexually) and engage in sexual activity (gene exchange), but what Talia is talking about in this post is the (arguably scientifically inaccurate) associations laypeople have with the term *asexual* after learning it in a science classroom.

2. In true patriarchal fashion, the "donor" cell was originally designated "male" and the recipient "female" (Bivins 2000).

3. In "Do Bacteria Have Sex?" (2001), Rosemary J. Redfield critiques many scientific explanations for the evolution of lateral gene transfer. According to her, many scientists have interpreted the large number of transferred genes in modern bacterial genomes as evidence that gene transfer must be adaptive. However, she replies, because natural selection eliminates almost all deleterious changes, the

number of transferred genes doesn't itself demonstrate that gene transfer overall is beneficial or adaptive (634).

4. Not including the genes transferred long ago from mitochondria and chloroplast ancestors to the host genome. The theory of endosymbiosis posits that the mitochondria, chloroplasts, and possibly other organelles of eukaryotic cells were once independent bacteria cells. These were ingested by host cells, survived, and eventually fully incorporated. Some of the DNA of the ingested bacteria may have been incorporated into the host's genome, although mitochondria, for example, also retain their own DNA (M. Gray 2017).

5. When haploid cells from the same "parent" cell fuse, this is called intragametophytic selfing (which occurs in some ferns). When haploid cells from the same organism but not the same parent cell fuse, this is called selfing (e.g., when a hermaphroditic plant pollenates itself; selfing can occur in plants, mollusks, nematodes, and other species) (Glémin et al. 2019; Hedrick 1987).

6. The question of "what is an amoeba" is somewhat complex. Generally, amoebas are single-celled eukaryotic organisms (protistans) that can alter their shape by extending arm-like protrusions call pseudopodia. Amoebas used to be grouped together in a macrotaxon named Sarcodina. However, molecular analysis revealed that Sarcodina was not a monophyletic group. As a result, amoebas were reclassified among many other high-level taxonomic groups ("Amoeba" 2020; Pawlowski and Burki 2009). Researchers have found evidence of "meiotic genes" in amoebas and other asexually reproducing protists, leading them to conclude that the last eukaryotic common ancestor (LECA) was capable of full meiotic sex (Speijer et al. 2015; Goodenough and Heitman 2014; Hofstatter and Lahr 2019; Khan and Siddiqui 2015).

7. Mating types are sometimes described as equivalent to "sexes" (as in male and female), although this comparison is problematic (Sandford 2019). Mating types occur in some protistans and fungi. Organisms of the same mating type are the "same" and cannot engage in sexual reproduction; those of different mating types can. Some species have two mating types, while others have multiple types (Billiard et al. 2012).

8. Fungi are highly diverse in terms of life cycle and reproduction. There is a range between fungi that only reproduce asexually and that frequently reproduce sexually. Fungi vary in terms of selfing vs. outcrossing and in terms of number and kind of mating types. Fungi also vary in terms of how long they spend in the haploid state and how long they remain polykaryotic (Nieuwenhuis and James 2016; Watkinson et al. 2015).

9. Confusingly, sometimes the sporophyte stage (2n) is called "the sexual stage," while the gametophyte stage (n) is called "the asexual stage."

10. Double fertilization is a unique process that occurs in angiosperms (flowering plants). In double fertilization, one sperm fuses with the egg to form the zygote, while another fuses with two "polar nuclei" to form the endosperm (the food-storage tissue) of the seed (Raghavan 2005).

11. Androgenesis is a rare form of asexual or parasexual reproduction in which the male's gamete develops to produce a new male, without genetic contribution from a female. Most commonly, an egg and a sperm fuse, and then the genetic contributions of the egg are eliminated. Androgenesis occurs in a type of conifer and some types of clams, ants, and stick insects (Schwander and Oldroyd 2016).

12. Some parthenogenic vertebrates (mostly lizards) reproduce without any need for males or copulation at all. In some species (including fish and amphibians), the female's genetically unreduced egg only begins to develop into an organism after it is "poked" by a sperm (which makes no genetic contribution), a process called gynogenesis. In some species, there exists a parasexual process called hybridgenesis, in which the diploid (2n) female uses a special type of meiosis to produce a haploid (n) egg that only contains genes from its own "mother," not its "father." This egg is fertilized by a sperm to produce a diploid (2n) zygote, which grows into a diploid (2n) female organism. When mature, the new female organism can again produce an egg containing only genes from its "mother." This form of reproduction could be classified as either sexual or asexual (Avise 2008). Finally, researchers have also discovered kleptogenesis, which is a type of gynogenesis or hybridgenesis but with the occasional incorporation of sperm-derived DNA into an otherwise asexual lineage (Avise 2015).

13. The concept of species is typically dependent on sexual reproduction, as species are often defined as a population of organisms capable of producing fertile offspring with each other. Obviously, this concept does not translate easily for asexually reproducing organisms. Scientists often use the term *clonal lineage* or *unisexual biotype* to indicate a group of clones deriving from a common ancestor (Avise 2015). Some claim that the concept of species can still be applied to asexually reproducing organisms, as there may be groups characterized by different morphologies and that behave, essentially, like species (Fontaneto and Barraclough 2015).

14. In mammals, parthenogenesis is limited because of problems arising from genomic imprinting, which, in this case, refers to the fact that males and females produce different epigenetic modifications of the genome during gamete (sperm/egg) formation, which leads to a different expression of imprinted genes from the maternal and paternal alleles. In order for viable development, an embryo needs this differential expression of imprinted genes (Kono 2006). Researchers have long sought to induce parthenogenesis in the lab. In the 1930s and 1940s, American biologist Gregory Pincus claimed he was able to induce parthenogenesis among rabbits, but his results have not been repeated. In the early 2000s, Tomohiro Kono and colleagues were able to genetically modify a female mouse such that her egg could be combined with one from an unmodified mouse to produce at least one viable offspring. In some cases, this process was referred to as parthenogenesis, and some hoped it would increase the suite of assisted reproductive technologies available to same-sex couples. Eva Mae Gillis-Buck (2016) discusses some of the

contradictory language used by scientists and the media in regard to Kono's research, although she seems to employ the term *parthenogenesis* to mean "female-only" reproduction rather than development of an unfertilized egg. For the purposes of this chapter, these experiments by Kono would not count as asexual or parthenogenic, as the mouse offspring combined the genetic information of two (female) individuals and did not result from an unfertilized egg. In 2022, researchers reported that they were able to generate viable full-term offspring directly from single unfertilized mouse eggs—which fits the definition of asexual reproduction and parthenogenesis (Y. Wei et al. 2022).

15. The three groups thought to be "ancient" asexuals are bdelloid rotifers, darwinulid ostracods, and oribatid mites (Fontaneto and Barraclough 2015). Additional groups for which there is evidence of long-lasting asexuality include stick insects (Schwander et al. 2011) and a type of roundworm (Fradin et al. 2017), among others.

16. The costs of sex include the cost of meiosis (which is a more time- and energy-intensive process than mitosis), those associated with finding a mate, and the costs (or really risks) that come from breaking up the parental genotype (which must have worked in that environment fairly well). This last cost is called the recombination load. If the species produces two forms of gametes (called male and female), the cost of sex can increase even further (the twofold cost only applies to sexually dimorphic species). Many scientists point out that it is an exaggeration to say that sex has a "twofold cost." The "cost of males" can be reduced if the species has more females than males and/or if males contribute resources to producing or caring for offspring (Lehtonen et al. 2012). Meirmans et al. (2012) argue that the costs of sex in the real world vary based on constraints on the evolution of asexuality, ecological differentiation, and certain life-history traits. One of the costs of sex frequently mentioned is genome dilution—the claim that a sexually reproducing organism only passes on 50 percent of its genome. Lehtonen et al. (2012, 172) explain that this is not an actual cost of sex, because "only the genes that determine the mode of reproduction matter in this context."

17. According to Otto (2008), early models found that sexual reproduction would only be selected for if three conditions are met: (1) the population is under directional selection (meaning increased variation can improve the response to selection); (2) there is negative epistasis (or two mutations together lead to a less fit phenotype than expected from their effects when alone; in this case sexual reproduction can restore variation eliminated by past selection); and (3) epistasis is weak (if it is too strong, the cost of breaking apart beneficial genotypes is too great) (also Orive 2020).

18. For instance, if a model investigates what factors will lead a sexually reproducing dimorphic species to outcompete a parthenogenically reproducing species, it is attempting to answer the questions "why sexual reproduction?" and "why males?" together. If a model investigates what factors will lead a

sexually reproducing species with isogamy (equal-sized gametes) to outperform a parthenogenically reproducing species, it is attempting to answer the question "why sex?" only. Most scientists agree that isogamy came first, then evolved into anisogamy, possibly in several ancestors of modern plants and animals (Lehtonen et al. 2016). It is unclear whether the evolution of anisogamy led first to hermaphroditism (in which an organism produces male and female gametes) or first to gonochorism (in which an organism produces either male or female gametes) or if hermaphroditism and gonochorism evolved independently. For most animals, gonochorism is thought to have evolved from hermaphroditism (Xie et al. 2021). However, researchers have found evidence that sequential hermaphroditism evolved from gonochorism in several species of fish (Sunobe et al. 2017). For angiosperms, gonochorism or dioecy is thought to have evolved from simultaneous hermaphroditism (Leonard 2018).

19. Other theories for the initial development of sexual reproduction include an infection model, according to which transposable elements may have arisen that were capable of making their host cells reproduce sexually with partners, allowing for their own propagation throughout a population. Another theory is related to the process of meiosis itself, maintaining that the proper sorting of chromosomes during meiosis is dependent on the formation of chiasmata (the points of contact between homologous chromosomes) and recombination was/is an (accidental) by-product of this system for sorting chromosomes. However, all three of these theories (repair, infection, and sorting) have significant limitations, and most scientists believe that they are not capable of explaining the maintenance of sex over time (Orive 2020; Otto and Lenormand 2002; Hartfield and Keightley 2012).

20. As an example of an explanation not covered here, Geoffrey E. Hill (2019) contends that the primary benefit of sexual reproduction is to allow a species to compensate for the mutational erosion of mitochondrial genes. As an example of a different organization scheme, in a review, Hartfield and Keightley (2012) organize theories into three categories: (1) breaking apart selection interference, including Hill–Robertson interference, the Fisher–Muller hypothesis, and Muller's ratchet; (2) parasitic resistance/the Red Queen hypothesis; and (3) mutational deterministic/Kondrashov's hatchet.

21. An environment might vary over time due to biotic factors (e.g., parasites or predation) or abiotic factors (e.g., changing weather patterns). Several theories have related benefits of sexual reproduction to an environment varying over space; for example, the "tangled bank hypothesis" argues that sexual offspring may be better able to exploit a variety of microsites near their parents compared to the clones of an asexually reproducing organism (Song et al. 2011; Becks and Agrawal 2010).

22. To give an overly simplified example of how this works in terms of genetic recombination, in a parasitic relationship, members of the host species might have two types of resistance: resistance 1 (coded by the genotypes AB or ab) and

resistance 2 (coded by the genotypes Ab or aB). The parasite species might have a way to overcome resistance 1 (coded by the genotypes CD or cd) and a way to overcome resistance 2 (coded by the genotypes Cd or cD). If in one generation, resistance 1 (AB or ab) is more common in the host species, the way to overcome it (CD or cd) will spread in the parasite species over a few generations. In response, by breaking up and recombining genotypes, sexual reproduction can allow resistance 2 (Ab or aB) to spread in the host species over a few generations. In response to this, sexual reproduction can allow the way to overcome it (Cd or cD) to spread in the parasite species over a few generations. In response to this, sexual reproduction can again allow resistance 1 (AB or ab) to spread in the host species over a few generations, and so on and so forth in a "flip-flopping" pattern (Michod 1996).

23. Feminist and queer critics have produced an enormous body of scholarship about reproduction, documenting the ways in which white, middle-class women are often pressured to biologically reproduce, while poor people, people of color, and sexual and gender minorities are often deterred or prevented from doing so ("stratified reproduction") (Agigian 2007; McCormack 2005). Feminists have also maintained that the differential responsibility of women and other minoritized groups for social reproduction (including biological reproduction and the activities required to keep people alive, such as cooking, cleaning, childcare, and eldercare) partially explains gender and other forms of inequality (Rodríguez-Rocha 2021). Academics have also analyzed efforts by minoritized people to reproduce despite societal prohibitions. Some of these efforts have included biological reproduction, such as the use by queer people of assisted reproductive technologies (including in vitro fertilization and surrogacy) (Mamo 2007; Mamo and Alston-Stepnitz 2015). Some of these efforts have included nonbiological forms of reproduction, such as queer kinship or families of choice (Weston 1997) and othermothering (Collins 2005; Stanlie James 1993). In addition, in queer theory, there is a strand of writing that rejects reproduction (Edelman 2004). The question of asexual reproduction does not fit neatly into any of these debates, as both sexual and asexual reproduction are forms of biological reproduction. In the first, genetic materials from more than one organism are brought together in the offspring; in the second, (in theory), the offspring is genetically identical to the "parent." As mentioned in note 14, there has been some effort to recruit asexual reproduction into the suite of assisted reproductive technologies that could allow people to biologically reproduce outside of a monogamous heterosexual couple. However, until recently, scientists did not think it was possible for mammals to reproduce asexually. As science progresses in this area, it is possible that in the future, asexual reproduction will be available as an assisted reproductive technology (Gillis-Buck 2016). If this is the case, it will likely fit into systems of stratified reproduction in ways similar to other assisted reproductive technologies (i.e., primarily available to "assimilated," wealthy, white people) (Mamo and Alston-Stepnitz 2015).

24. But, for a discussion of sexual selection in hermaphrodites, see Beekman et al. 2016; Lorenzi and Sella 2008.
25. Weismann is best known for his theory of germ plasm, positing that in multicellular organisms, only specialized cells (germ cells) can pass genetic information on to offspring, while other cells (somatic cells) cannot. According to Weismann, the germ plasm is unchanging, and thus acquired variation cannot be passed on. As historian of science Stephanie Meirmans (2009) explains, Weismann needed some way to account for the creation of the variation necessary for natural selection if the germ line remains unchanging. Sex was his solution to the problem, as he saw it as creating genetic variation. One source states that Weismann was one of the honorary vice presidents of the first Eugenics Congress, but I did not find any more information about his relationship to eugenics in the secondary literature (Blom 2008). However, Weismann's arguments against the inheritance of acquired variation were important for early eugenicists (Larson 2010; Sussman 2016; Cowan 2016). Ronald A. Fisher, one of the most influential evolutionary biologists of the twentieth century, was a committed eugenicist. Meirmans (2009) suggests that Fisher was motivated to identify an evolutionary explanation for sexual reproduction because he wanted to reconcile Darwin's theory of natural selection with Mendelian genetics—he claimed that natural selection would lead to a Mendelian system of inheritance through sex, because this would allow for the constant generation of variation without the need for a high level of mutation (it was understood, by Fisher's time, that most mutations are deleterious). Fisher's scientific work in evolutionary biology—including on sexual reproduction—was in the service of his eugenic thinking; for instance, one-third of his most famous book, *The Genetical Theory of Natural Selection*, was devoted to eugenics. He was concerned with the fecundity of the poor compared to the elite in England and advocated for a voluntary eugenic sterilization law. He defended a Nazi eugenicist, expressed concerns about miscegenation, and was one of a handful of scientists who publicly objected to UNESCO's post–World War II statement on race, because he believed that human groups differ "in their innate capacity for intellectual and emotional development," although defenders hold that he did not support racial discrimination (Bodmer et al. 2021; E. Johnson 2021). Herman J. Muller was also a eugenicist, but a socialist one. In the 1930s, he articulated ideas similar to Fisher's about the evolutionary importance of sexual reproduction for generating variation (which is why this theory became known as the Fisher-Muller hypothesis). However, most of his research was focused on the role of radiation in producing (usually deleterious) genetic mutations, which he used to advocate against nuclear weapons testing. Muller proposed the theory that became known as "Muller's ratchet" in 1964, contending that without sexual reproduction, deleterious mutations build up irreversibly over time in small populations. Thus, for Muller, sexual reproduction was

advantageous for its ability to limit their effects. Politically, Muller combined his socialist and eugenic ideals by arguing for the need to improve both human environments and human genotypes. Later in life, he advocated for a program of positive eugenics called "Voluntary Choice," involving the creation of sperm banks containing only that of "genetically superior men," which could be employed by people already pursuing artificial insemination. In the distant future, he hoped, perhaps all who planned to procreate would choose artificial insemination using the sperm of genetically superior men (Crow 2005; Carlson 1981; Allen 1970). By the 1960s and 1970s, evolutionary biology had dissociated itself from eugenics, which was no longer a widely accepted political ideology. John Maynard Smith and George C. Williams were both members of the political left; Maynard Smith in particular was a member of the Communist Party for many years, although he became disillusioned with it eventually (Charlesworth 2004; Piel 2017; Futuyma and Stearns 2010; Meyer 2010). Still, as mentioned, their scientific work may have contributed to a deterministic view of genes.

26. I vaguely recall having this reaction myself when encountering discussions about the paradox of sex during an introductory biology course in my first year of college.

27. Citations are from Hiebert et al. (2021): J. B. C. Jackson, "Competition on Marine Hard Substrata: The Adaptive Significance of Solitary and Colonial Strategies," *American Naturalist* 111 (1977), 743–67; J. L. Harper, "Modules, Branches, and the Capture of Resources," in *Population Biology and Evolution of Clonal Organisms,* edited by J. B. C. Jackson, L. W. Buss, and R. E. Cook, 1–33. (Yale University Press, 1985).

28. Like many others, Margulis contends that lateral gene transfer evolved in bacteria in order to repair DNA. However, her explanation for the development of sexual reproduction in eukaryotes is unique. According to Margulis (and her coauthor and son Dorian Sagan), diploidy (2n) arose as a result of starvation and cannibalism—in times of scarcity, cells consumed other cells without fully digesting them, ending up with two sets of genes. As conditions improved, diploid (2n) cells that were able to divide back into haploid (n) cells through meiosis were more likely to survive. When conditions of scarcity arose again, these haploid (n) cells fused, and so on and so forth, leading to a regular cycle of fusion and meiotic cell division. At this point, Margulis's argument becomes less clear to me. She and Sagan seem to claim simultaneously that meiosis persists because (1) once it evolved, eukaryotic cells were essentially stuck with it, and/or (2) meiosis plays an important role in multicellular organisms with somatic differentiation, namely: morphological complexity requires a complex genome, this complex genome needs to be regularly checked and/or adjusted, and this checking or correcting occurs (somehow) through meiosis (Margulis and Sagan 1990). Others have also found this part of the argument confusing (Bell 1988; Ayala 1989).

29. Other examples include *The Female Man* (1975) by Joanna Russ, *Motherlines*

(1978) by Suzy McKee Charnas, *A Door into Ocean* (1986) by Joan Slonczewski, and *Girl One: A Novel* (2021) by Sara Flannery Murphy, among others.

30. For discussions of reproduction in feminist science fiction, see Bonnevier 2023; Gilarek 2013; B. Nichols 2020. For discussions of racism and eugenics in *Herland*, see Egan 2011; Seitler 2003; Weinbaum 2001. For various perspectives on the politics of women-only separatist spaces, see Frye 1997; Enszer 2016; Browne 2009; Rudy 2001.

31. Lynn Margulis (1938–2011) was an evolutionary biologist best known for her theory of symbiogenesis, or the idea that many important evolutionary advances have come about through endosymbiosis. Specifically, she posited that the mitochondria, chloroplasts, and flagella or cilia of eukaryotic cells were once independent bacteria cells. Today, it is widely accepted that mitochondria and chloroplasts did arise through endosymbiosis, but there remains no evidence that flagella or cilia are of endosymbiotic origin (M. Gray 2017). Margulis and her work have been beloved by feminist thinkers for many reasons: She was a woman advocating for an arguably "feminine" theory (symbiosis, which involves cooperation and mutual benefit, easily reads as feminine) against the neo-Darwinian consensus of a male-dominated profession; and her research was initially rejected by a number of (male) scientists, only to be (at least) partially confirmed through later research (Horgan 2011; Lake 2011). Margulis's insistence on the importance of symbiosis is itself appealing to feminist scholars, as her research provides support for the productivity of collectivity and connection while also undermining self/other dichotomies (Basile 2021; Lopez 2001; Subramaniam 2016). As discussed elsewhere, however, her arguments about sexual reproduction are counter to the current scientific consensus and remain unsubstantiated.

32. I admit that I may not fully capture the nuances of Parisi's argument. She extensively references theorists such as Bergson, Lucretius, and Spinoza, whose writing I am not thoroughly familiar with. In a review, Stella Sandford (2004) describes *Abstract Sex* as a "hellish read" due to its convoluted reasoning and language. For another critique of the "molecularization of sexuality," see J. Rosenberg 2014.

33. Some justice projects do not aim to transform human societies; for example, Kirksey (2019) turns to *Wolbachia* as a model for how to "love in a time of extinction," including the possibility of human extinction.

34. Of course, as has already been discussed, prokaryotes reproduce asexually, and asexual reproduction persists at many levels for multicellular eukaryotic organisms that reproduce sexually at the level of the organism: for instance, soma cells reproduce asexually through mitosis. Mitochondria also reproduce asexually. For a discussion of the origin and persistence of the mitochondrial genome, see Jansen 2000.

35. The question of whether sexual reproduction is "necessary" for complex, multicellular organisms hinges on how one defines "necessary." Clearly, sexual

reproduction is not entirely necessary, as some complex organisms are able to reproduce parthenogenically. The counterargument is that asexually reproducing multicellular "species" are confined to specific environments, tend not to survive long in evolutionary terms, and/or do not, themselves, lead to species diversification, due to limiting factors. A counter-counterargument is that, theoretically, there could be other ways for species to overcome these limiting factors, and there is evidence that asexually reproducing species have found ways to do so. However, it has also been claimed that, given starting constraints, sexual reproduction was the best or only option to do so. For example, Michod (1996) holds that more complex organisms required much larger genomes (which may then have needed to be packaged into nuclei) and that the replication error rate for such a genome required some kind of additional repair mechanism, and the only (or best) solution to this for a nucleated cell was developing diploidy (keeping two sets of genes in the nucleus). But then diploidy led to the issue of accumulating deleterious mutations (which can be "masked" by heterozygosity in diploid organisms), and this, in turn, necessitated outcrossing, or sexual reproduction of some kind. Whether one accepts Michod's specific argument, it still seems risky to base a political argument on the claim that sexual reproduction is not necessary.

36. Roberta Bivins (2000) offers an excellent in-depth discussion of the imposition of a concept of heterosexual sex on bacterial lateral gene transfer.

37. For instance, in the article "Connecting Theory and Data to Understand Recombination Rate Evolution," the authors seem to be using recombination to mean only crossing over (Dapper and Payseur 2017). However, the entry titled "Recombination and Sex" in *The Princeton Guide to Evolution* defines recombination as crossing over plus independent assortment (Barton 2013).

38. For a discussion about different meanings of the word *clone* in scientific research, see Martens et al. 2009.

39. For a further discussion of the genetic mechanisms involved in different types of parthenogenesis, see Jaron et al. 2021.

40. Although, according to Aanen et al. (2016), some scientists call mobile gametes "male" and stationary gametes "female."

Conclusion

1. For a discussion of Foucault's rejection of "sex-desire" in favor of "bodies and pleasures," see Butler 1999; McWhorter 1997; Sawicki 2010.

2. See also Glick 2000 for a discussion of how some sex-positive feminist (and queer) theories and activism in the 1980s and 1990s tended to essentialize sexual desire and activity.

3. This example appears in an earlier publication (Gupta 2022).

4. In his article "Taxonomically Queer? Sexology and New Queer, Trans, and Asexual Identities" (2023), Kadji Amin analyzes the seemingly endless

proliferation of microidentity categories in queer, trans, and asexual communities. He notes the similarities between these contemporary taxonomical efforts and sexologist Magnus Hirschfeld's 1910 "theory of intermediaries." Amin simultaneously recognizes the creativity of this taxonomical effort ("It embodies utopian hopes for a world in which no one's gender, sexuality, or mode of attraction would be presumed in advance and in which everyone would have recourse to a nuanced and nimble vocabulary through which to know, define, and communicate their own unique gender and sexual subjectivity" [92]), while also acknowledging the historical and contemporary imbrication of taxonomical projects with colonial and anti-Black projects of power and knowledge: "New queer taxonomies end up dovetailing, sometimes intentionally (Chasin 2013), with popular and medical beliefs that sexual and gender identities are biologically rooted, innate, unchanging, and discrete from one another . . . new queer classification systems use taxonomy to make an implicit bid for the rationality, legibility, and scientific validity of otherwise quirky outsider identities" (100).

5. For a discussion of masturbation by nonhuman animals, see Roth et al. 2023.

6. Defining sex based on orgasm is even more problematic. The scientific and feminist/queer literature on the topic is extensive, and it is beyond the scope of this chapter or book to summarize either. In short, there remains significant controversy about the evolutionary "purpose" of female orgasm (Lloyd 2005). There is also a substantial body of feminist and queer research critiquing the "orgasmic imperative" (or pressure to experience orgasm) (Bosley 2010; Duncombe and Marsden 1996; Willis et al., 2018; Gerhard 2000; Tuana 2004; Musser 2012; Jagose 2012). For an asexuality studies discussion of orgasm, see Cerankowski 2021.

7. Rereading the scientific literature on asexual and sexual reproduction can also destabilize the concept of sexual desire. Researchers argue that bacteria engage in sex (as in genetic combination) separate from reproduction, and that fungi and plants engage in sex (as in genetic combination) as part of reproduction. But do bacteria experience the desire to engage in sex (genetic combination)? Do fungi and plants experience the desire to engage in sex (sexual reproduction)? It seems unlikely that these organisms experience subjective desire. Is the concept limited then to the desire for copulation or mating in human and nonhuman animals?

8. The practice of "queer reading," in particular, seems to expand the domain of the sexual, given that its purpose is often to make visible the invisible in texts—sometimes the operation of heterosexism but also those of queer sexual desires and practices (Björklund and Lönngren 2020).

9. As Ruth Ginzberg (1992) points out, Lorde may not be defining the ancient Greek meaning of "eros" correctly. There were at least three distinct conceptions of "love" in ancient Greece: agape, eros, and philia. Bennett Helm (2009) defines these as follows: "Agape is a kind of love that does not respond to

the antecedent value of its object but instead is thought to create value in the beloved; it has come through the Christian tradition to mean the sort of love God has for us persons as well as, by extension, our love for God and our love for humankind in general. By contrast, eros and philia are generally understood to be responsive to the merits of their objects—to the beloved's properties, especially his goodness or beauty. The difference is that eros is a kind of passionate desire for an object, typically sexual in nature, whereas 'philia' originally meant a kind of affectionate regard or friendly feeling towards not just one's friends but also possibly towards family members, business partners, and one's country at large." Merriam-Webster's dictionary provides two definitions of *erotic*: "of, devoted to, or tending to arouse sexual love or desire" and "strongly marked or affected by sexual desire" ("Definition of EROTIC," n.d.). This is not to say that the term is used only or always to mean sexual desire exclusively but rather that sexual desire is an embedded connotation.

REFERENCES

Aanen, Duur, Madeleine Beekman, and Hanna Kokko. 2016. "Weird Sex: The Underappreciated Diversity of Sexual Reproduction." *Philosophical Transactions of the Royal Society B: Biological Sciences* 371 (1706): 20160262.

Abbott, Elizabeth. 2000. *A History of Celibacy*. Simon & Schuster.

Abler, Birgit, Daniela Kumpfmüller, Georg Grön, Martin Walter, Julia Stingl, and Angela Seeringer. 2013. "Neural Correlates of Erotic Stimulation Under Different Levels of Female Sexual Hormones." *PLOS ONE* 8 (2): e54447.

Ace and Aro Advocacy Project. 2022. "Aspecs and Queer Platonic Relationships—Part One." July 16. https://taaap.org/2022/07/16/qprs-part-one/.

"Ace/Aro Mythbusting." n.d. Oxford University LGBTQ+ Society. Accessed May 1, 2023. https://www.oulgbtq.org/acearo-mythbusting.html.

Adriaens, Pieter R., and Andreas De Block. 2022. *Of Maybugs and Men: A History and Philosophy of the Sciences of Homosexuality*. University of Chicago Press.

Agigian, Amy. 2007. "Stratified Reproduction." In *The Blackwell Encyclopedia of Sociology*, edited by George Ritzer, 4835–37. Blackwell.

Ågmo, Anders. 1976. "Individual Differences in the Response to Androgen in Male Rabbits." *Physiology & Behavior* 17 (4): 587–89.

Ågmo, Anders. 1999. "Sexual Motivation—An Inquiry into Events Determining the Occurrence of Sexual Behavior." *Behavioural Brain Research* 105 (1): 129–50.

Ågmo, Anders, and Ellen Laan. 2022. "Sexual Incentive Motivation, Sexual Behavior, and General Arousal: Do Rats and Humans Tell the Same Story?" *Neuroscience & Biobehavioral Reviews* 135 (April): 104595.

Ågren, J. Arvid. 2016. "Selfish Genetic Elements and the Gene's-Eye View of Evolution." *Current Zoology* 62 (6): 659–65.

Alaimo, Stacy. 2016. "Eluding Capture: The Science, Culture, and Pleasure of 'Queer' Animals." Chap. 2 in *Exposed: Environmental Politics and Pleasures in Posthuman Times*. University of Minnesota Press.

Alarcão, Soraia M., and Manuel J. Fonseca. 2018. "Identifying Emotions in Images from Valence and Arousal Ratings." *Multimedia Tools and Applications* 77 (13): 17413–35.

Allen, Garland E. 1970. "Biology and Culture: Science and Society in the Eugenic Thought of H. J. Muller." *BioScience* 20 (6): 346–53.

Allison, Katie. 2015. "So You Reproduce like a Starfish?" *The Stranger*, June 24. https://www.thestranger.com/features/2015/06/24/22437200/so-you-reproduce-like-a-starfish.

Altimus, Cara M., Bianca Jones Marlin, Naomi Ekavi Charalambakis, et al. 2020. "The Next 50 Years of Neuroscience." *Journal of Neuroscience* 40 (1): 101–6.

Altman, Mara. 2010. "Rutgers Lab Studies Female Orgasm Through Brain Imaging." NJ.com, April 20. https://www.nj.com/insidejersey/2010/04/science_consciousness_and_the.html.

Ambur, Ole Herman, Jan Engelstädter, Pål J. Johnsen, Eric L. Miller, and Daniel E. Rozen. 2016. "Steady at the Wheel: Conservative Sex and the Benefits of Bacterial Transformation." *Philosophical Transactions of the Royal Society B: Biological Sciences* 371 (1706): 20150528.

American Psychiatric Association. 1952. *Diagnostic and Statistical Manual of Mental Disorders*. American Psychiatric Association Mental Hospital Service.

American Psychiatric Association. 1968. *Diagnostic and Statistical Manual of Mental Disorders, Second Edition (DSM-II)*. American Psychiatric Association.

American Psychiatric Association. 1981. *Diagnostic and Statistical Manual of Mental Disorders, Third Edition (DSM-III)*. American Psychiatric Association.

American Psychiatric Association. 1987. *Diagnostic and Statistical Manual of Mental Disorders, Third Edition Revised (DSM-III-R)*. American Psychiatric Association.

American Psychiatric Association. 1994. *Diagnostic and Statistical Manual of Mental Disorders, Fourth Edition (DSM-IV)*. American Psychiatric Publishing.

American Psychiatric Association. 2013. *Diagnostic and Statistical Manual of Mental Disorders, Fifth Edition: DSM-5*. American Psychiatric Association.

Amin, Kadji. 2023. "Taxonomically Queer? Sexology and New Queer, Trans, and Asexual Identities." *GLQ: A Journal of Lesbian and Gay Studies* 29 (1): 91–107.

"Amoeba." 2020. *The Gale Encyclopedia of Science* 1 (November): 189–90.

Antonio-Cabrera, Edwards, and Raúl G. Paredes. 2014. "Testosterone or Oestradiol Implants in the Medial Preoptic Area Induce Mating in Noncopulating Male Rats." *Journal of Neuroendocrinology* 26 (7): 448–58.

Antonsen, Amy N., Bozena Zdaniuk, Morag Yule, and Lori A. Brotto. 2020. "Ace and Aro: Understanding Differences in Romantic Attractions Among Persons Identifying as Asexual." *Archives of Sexual Behavior* 49 (5): 1615–30.

Arnow, Bruce A., John E. Desmond, Linda L. Banner, et al. 2002. "Brain Activation and Sexual Arousal in Healthy, Heterosexual Males." *Brain* 125 (5): 1014–23.

Arnow, Bruce A., Leah S. Millheiser, Amy S. Garrett, et al. 2009. "Women with Hypoactive Sexual Desire Disorder Compared to Normal Females: A Functional Magnetic Resonance Imaging Study." *Neuroscience* 158 (2): 484–502.

Attwood, Feona, ed. 2009. *Mainstreaming Sex: The Sexualisation of Western Culture*. I. B. Tauris.

AVEN: The Asexuality Visibility and Education Network. n.d.-a. "AVEN Survey 2008—Results." Accessed August 15, 2011. http://www.asexuality.org/home/2008_stats.html.

AVEN: The Asexuality Visibility and Education Network. n.d.-b. "Pandas as Asexual Symbol." Asexual Visibility and Education Network. Accessed December 13, 2022. https://www.asexuality.org/en/topic/200477-pandas-as-asexual-symbol/.

AVEN: The Asexuality Visibility and Education Network. n.d.-c. "Pandas Should Be a (Gray-)Asexual Symbol." Asexual Visibility and Education Network. Accessed December 13, 2022. https://www.asexuality.org/en/topic/141808-pandas-should -be-a-gray-asexual-symbol/.

Avise, John C. 2008. *Clonality: The Genetics, Ecology, and Evolution of Sexual Abstinence in Vertebrate Animals.* Oxford University Press.

Avise, John C. 2015. "Evolutionary Perspectives on Clonal Reproduction in Vertebrate Animals." *Proceedings of the National Academy of Sciences* 112 (29): 8867–73.

Ayala, Francisco J. 1989. "Origins of Sex: Three Billion Years of Genetic Recombination." *BioScience* 39 (1): 45–47.

Bagemihl, Bruce. 2000. *Biological Exuberance: Animal Homosexuality and Natural Diversity.* St. Martin's Publishing Group.

Baid, Rashmi, and Rakesh Agarwal. 2018. "Flibanserin: A Controversial Drug for Female Hypoactive Sexual Desire Disorder." *Industrial Psychiatry Journal* 27 (1): 154–57.

Bailey, J. Michael, Paul L. Vasey, Lisa M. Diamond, S. Marc Breedlove, Eric Vilain, and Marc Epprecht. 2016. "Sexual Orientation, Controversy, and Science." *Psychological Science in the Public Interest* 17 (2): 45–101.

Bakker, Julie. 2019. "The Sexual Differentiation of the Human Brain: Role of Sex Hormones Versus Sex Chromosomes." In *Neuroendocrine Regulation of Behavior*, edited by Lique M. Coolen and David R. Grattan, 45–67. Springer International Publishing.

Balcombe, Jonathan. 2009. "Animal Pleasure and Its Moral Significance." *Sentience Collection*, May. https://www.wellbeingintlstudiesrepository.org/acwp_asie/5.

Bancroft, John, Cynthia A. Graham, Erick Janssen, and Stephanie A. Sanders. 2009. "The Dual Control Model: Current Status and Future Directions." *Journal of Sex Research* 46 (2–3): 121–42.

Baniel, Alice, Guy Cowlishaw, and Elise Huchard. 2017. "Male Violence and Sexual Intimidation in a Wild Primate Society." *Current Biology* 27 (14): 2163–2168.e3.

Barad, Karen. 2007. *Meeting the Universe Halfway: Quantum Physics and the Entanglement of Matter and Meaning.* Duke University Press.

Barazandeh, Marjan, Corey S. Davis, Christopher J. Neufeld, David W. Coltman, and A. Richard Palmer. 2013. "Something Darwin Didn't Know About Barnacles: Spermcast Mating in a Common Stalked Species." *Proceedings of the Royal Society B: Biological Sciences* 280 (1754): 20122919.

Barker, Meg-John, and Justin Hancock. 2019. "Asexuality and Trauma." *Meg-John & Justin* (blog), November 15. https://megjohnandjustin.com/you/asexuality -and-trauma/.

Barounis, Cynthia. 2014. "Compulsory Sexuality and Feminist/Crip Resistance in John Cameron Mitchell's *Shortbus*." In *Asexualities: Feminist and Queer Perspectives*, edited by K. J. Cerankowski and Megan Milks, 174–96. Routledge.

Bartels, Andreas, and Semir Zeki. 2000. "The Neural Basis of Romantic Love." *NeuroReport* 11 (17): 3829–34.

Barton, Nick H. 2013. "Recombination and Sex." In *The Princeton Guide to Evolution*, edited by Jonathan B. Losos, 328–33. Princeton University Press.

Barua, Maan. 2020. "Affective Economies, Pandas, and the Atmospheric Politics of Lively Capital." *Transactions of the Institute of British Geographers* 45 (3): 678–92.

Basile, Jonathan. 2021. "Symbioautothanatosis: Science as Symbiont in the Work of Lynn Margulis." *Síntesis: Revista de filosofía* 4 (2): 60–86.

Baska, Maggie. 2023. "NYC Pride's First Asexual Grand Marshal Explains Why Including Ace People at Pride Is So Important." *PinkNews*, June 1. https://www .thepinknews.com/2023/06/01/yasmin-benoit-asexual-grand-marshal-nyc-pride/.

Basson, Rosemary. 2000. "The Female Sexual Response: A Different Model." *Journal of Sex & Marital Therapy* 26 (1): 51–65.

Basson, Rosemary. 2001. "Using a Different Model for Female Sexual Response to Address Women's Problematic Low Sexual Desire." *Journal of Sex & Marital Therapy* 27 (5): 395–403.

Bateman, Angus J. 1948. "Intra-Sexual Selection in Drosophila." *Heredity* 2 (pt. 3): 349–68.

Batričević, Milica, and Andrej Cvetić. 2016. "Uncovering an A: Asexuality and Asexual Activism in Croatia and Serbia." In *Intersectionality and LGBT Activist Politics: Multiple Others in Croatia and Serbia*, edited by Bojan Bilić and Sanja Kajinić, 77–103. Palgrave Macmillan UK.

Bauer, Heike, ed. 2015. *Sexology and Translation: Cultural and Scientific Encounters Across the Modern World*. Temple University Press.

Beauvoir, Simone de. 2010. *The Second Sex*. Vintage Books.

Beck, Charles B. 2010. *An Introduction to Plant Structure and Development: Plant Anatomy for the Twenty-First Century*. Cambridge University Press.

Becks, Lutz, and Aneil F. Agrawal. 2010. "Higher Rates of Sex Evolve in Spatially Heterogeneous Environments." *Nature* 468 (7320): 89–92.

Bedos, Marie, Alfonso Longinos Muñoz, Agustín Orihuela, and José Alberto Delgadillo. 2016. "The Sexual Behavior of Male Goats Exposed to Long Days Is as Intense as During Their Breeding Season." *Applied Animal Behaviour Science* 184 (November): 35–40.

Beekman, Madeleine, Bart Nieuwenhuis, Daniel Ortiz-Barrientos, and Jonathan P. Evans. 2016. "Sexual Selection in Hermaphrodites, Sperm and Broadcast Spawners, Plants and Fungi." *Philosophical Transactions of the Royal Society B: Biological Sciences* 371 (1706): 20150541.

Bell, Graham. 1982. *The Masterpiece of Nature: The Evolution and Genetics of Sexuality*. University of California Press.

Bell, Graham. 1988. "Origins of Sex." *Journal of Genetics* 67 (1): 63–66.

Bem, Sandra Lipsitz. 1995. "Dismantling Gender Polarization and Compulsory Heterosexuality: Should We Turn the Volume Down or Up?" *Journal of Sex Research* 32 (4): 329–34.

Benedet, Janine, and Isabel Grant. 2014. "Sexual Assault and the Meaning of Power and Authority for Women with Mental Disabilities." *Feminist Legal Studies* 22 (2): 131–54.

Benjamin, Harry. 1966. *The Transsexual Phenomenon: A Scientific Report on Transsexualism and Sex Conversion in the Human Male and Female.* Julian Press.

Bennett, Judith M. 2000. "'Lesbian-like' and the Social History of Lesbianisms." *Journal of the History of Sexuality* 9 (1/2): 1–24.

Berglund, Hans, Per Lindström, and Ivanka Savic. 2006. "Brain Response to Putative Pheromones in Lesbian Women." *Proceedings of the National Academy of Sciences* 103 (21): 8269–74.

Bhat, Rashmi. 2018. "How Did an All-Female Species Survive sans Sexual Reproduction for Millennia?" *The Wire* (blog), February 19. https://thewire.in /science/female-species-survive-sans-sexual-reproduction-millennia.

Bianchi-Demicheli, Francesco, Yann Cojan, Lakshmi Waber, Nathalie Recordon, Patrik Vuilleumier, and Stephanie Ortigue. 2011. "Neural Bases of Hypoactive Sexual Desire Disorder in Women: An Event-Related fMRI Study." *Journal of Sexual Medicine* 8 (9): 2546–59.

Billiard, Sylvain, Manuela López-Villavicencio, Michael E. Hood, and Tatiana Giraud. 2012. "Sex, Outcrossing and Mating Types: Unsolved Questions in Fungi and Beyond." *Journal of Evolutionary Biology* 25 (6): 1020–38.

Birri, Marcela, Mariana Vallejo, Miguel Carro-Juárez, and A. Mariel Agnese. 2017. "Aphrodisiac Activity of *Phlegmariurus saururus* in Copulating and Noncopulating Male Rats." *Phytomedicine: International Journal of Phytotherapy and Phytopharmacology* 24 (January 15): 104–10.

Bivins, Roberta. 2000. "Sex Cells: Gender and the Language of Bacterial Genetics." *Journal of the History of Biology* 33 (1): 113–39.

Björklund, Jenny, and Ann-Sofie Lönngren. 2020. "Now You See It, Now You Don't: Queer Reading Strategies, Swedish Literature, and Historical (In)visibility." *Scandinavian Studies* 92 (2): 196–228.

Blanchard, Ray. 1989. "The Classification and Labeling of Nonhomosexual Gender Dysphorias." *Archives of Sexual Behavior* 18 (4): 315–34.

Blanchard, Ray. 2005. "Early History of the Concept of Autogynephilia." *Archives of Sexual Behavior* 34 (4): 439–46.

Blom, Philipp. 2008. *The Vertigo Years : Change and Culture in the West, 1900–1914.* McClelland & Stewart.

Boast, Hannah. 2022. "Theorizing the Gay Frog." *Environmental Humanities* 14 (3): 661–79.

Bodmer, Walter, R. A. Bailey, Brian Charlesworth, et al. 2021. "The Outstanding Scientist, R. A. Fisher: His Views on Eugenics and Race." *Heredity* 126 (4): 565–76.

Boislard, Marie-Aude, Stéfany Boisvert, Mélanie Millette, Laurence Dion, and Julie Lavigne. 2022. "Representations of Sexually Inexperienced Emerging Adults in Fictional Television Series and Movies." *Sexuality & Culture* 26 (3): 1031–59.

Boldgiv, Bazartseren. 2010. "George Christopher Williams (1926–2010): Gene's-Eye View of Evolution." *Mongolian Journal of Biological Sciences* 8 (2): 73–74.

Bonnevier, Jenny. 2023. "In the Womb of Utopia: Feminist Science Fiction, Repro- ductive Technology, and the Future." *American Studies in Scandinavia* 55 (1): 70–93.

Borg, Charmaine, Peter J. de Jong, and Janniko R. Georgiadis. 2014. "Subcortical BOLD Responses During Visual Sexual Stimulation Vary as a Function of Implicit Porn Associations in Women." *Social Cognitive and Affective Neuroscience* 9 (2): 158–66.

Borg, Charmaine, Janniko R. Georgiadis, Remco J. Renken, Symen K. Spoelstra, Willibrord Weijmar Schultz, and Peter J. de Jong. 2014. "Brain Processing of Visual Stimuli Representing Sexual Penetration Versus Core and Animal-Reminder Disgust in Women with Lifelong Vaginismus." *PLOS ONE* 9 (1): e84882.

Bosley, Jocelyn. 2010. "From Monkey Facts to Human Ideologies: Theorizing Female Orgasm in Human and Nonhuman Primates, 1967–1983." *Signs: Journal of Women in Culture and Society* 35 (3): 647–71.

Both, Stephanie, Walter Everaerd, and Ellen Laan. 2007. "Desire Emerges from Excitement: A Psychophysiological Perspective on Sexual Motivation." In *The Psychophysiology of Sex*, edited by Erick Janssen, 327–39. Indiana University Press.

Bradshaw, Julia, Natalie Brown, Alan Kingstone, and Lori Brotto. 2021. "Asexuality vs. Sexual Interest/Arousal Disorder: Examining Group Differences in Initial Attention to Sexual Stimuli." *PLOS ONE* 16 (12): e0261434.

Bradway, Teagan, and Elizabeth Freeman, eds. 2022. *Queer Kinship: Race, Sex, Belonging, Form*. Duke University Press.

Brandley, ben, and Marco Dehnert. 2023. "'I Am Not a Robot, I Am Asexual': A Qualitative Critique of Allonormative Discourses of Ace and Aro Folks as Robots, Aliens, Monsters." *Journal of Homosexuality* 71 (6): 1560–83.

Brandley, ben, and Leland G. Spencer. 2023. "Rhetorics of Allonormativity: The Case of Asexual Latter-Day Saints." *Southern Communication Journal* 88 (1): 1–15.

Brennan, Toni. 2015. "Eugenics and Sexology." In *The International Encyclopedia of Human Sexuality*, edited by Patricia Whelehan and Anne Bolin, 325–68. John Wiley & Sons.

Brotto, Lori A. 2010a. "The DSM Diagnostic Criteria for Hypoactive Sexual Desire Disorder in Men." *Journal of Sexual Medicine* 7 (6): 2015–30.

Brotto, Lori A. 2010b. "The DSM Diagnostic Criteria for Hypoactive Sexual Desire Disorder in Women." *Archives of Sexual Behavior* 39 (2): 221–39.

Brotto, Lori A., and K. B. Smith. 2014. "Sexual Desire and Pleasure." In *APA Handbook of Sexuality and Psychology*. Vol. 1, *Person-Based Approaches*, 205–44.

Brotto, Lori A., and Morag A. Yule. 2011. "Physiological and Subjective Sexual Arousal in Self-Identified Asexual Women." *Archives of Sexual Behavior* 40 (4): 699–712.

Brown, Gail. 2001. "Sexual vs. Asexual Reproduction: Scientists Find Sex Wins." *The Current UCSB* (blog). October 19, 2001. https://www.news.ucsb.edu/2001/011513 /sexual-vs-asexual-reproduction-scientists-find-sex-wins.

Brown, Natalie B., Diana Peragine, Doug P. VanderLaan, Alan Kingstone, and Lori A. Brotto. 2021. "Cognitive Processing of Sexual Cues in Asexual Individuals and Heterosexual Women with Desire/Arousal Difficulties." *PLOS ONE* 16 (5): e0251074.

Browne, Kath. 2009. "Womyn's Separatist Spaces: Rethinking Spaces of Difference and Exclusion." *Transactions of the Institute of British Geographers* 34 (4): 541–56.

Buchen, Lizzie. 2008. "Pandas: Evolution's Big Fat (Adorable) Mistake?" *NBC News*, August 5. https://www.nbcnews.com/id/wbna26036245.

Bühler, Mira, Sabine Vollstädt-Klein, Jane Klemen, and Michael N. Smolka. 2008. "Does Erotic Stimulus Presentation Design Affect Brain Activation Patterns? Event-Related vs. Blocked fMRI Designs." *Behavioral and Brain Functions* 4 (1): 30.

Bullard, David G. 1988. "The Treatment of Desire Disorders in the Medically Ill and Physically Disabled." In *Sexual Desire Disorders*, edited by Sandra R. Leiblum and Raymond C. Rosen, 348–84. Guilford Press.

Burchard, Johann M. 1965. "Psychopathology of Transvestism and Transsexualism." *Journal of Sex Research* 1 (1): 39–43.

Burke, Nathan W., and Russell Bonduriansky. 2017. "Sexual Conflict, Facultative Asexuality, and the True Paradox of Sex." *Trends in Ecology & Evolution* 32 (9): 646–52.

Buss, David M. 1998. "Sexual Strategies Theory: Historical Origins and Current Status." *Journal of Sex Research* 35 (1): 19–31.

Butler, Judith. 1999. "Revisiting Bodies and Pleasures." *Theory, Culture & Society* 16 (2): 11–20.

Cabaj, Robert. 2009. "Strike While the Iron Is Hot: Science, Social Forces and Ego-Dystonic Homosexuality." *Journal of Gay & Lesbian Mental Health* 13 (2): 87–93.

Cacchioni, Thea. 2015a. *Big Pharma, Women, and the Labour of Love.* University of Toronto Press.

Cacchioni, Thea. 2015b. "The Medicalization of Sexual Deviance, Reproduction, and Functioning." In *Handbook of the Sociology of Sexualities*, edited by John DeLamater and Rebecca F. Plante, 435–52. Springer.

Cacioppo, Stephanie, Francesco Bianchi-Demicheli, Chris Frum, James G. Pfaus, and James W. Lewis. 2012. "The Common Neural Bases Between Sexual Desire and Love: A Multilevel Kernel Density fMRI Analysis." *Journal of Sexual Medicine* 9 (4): 1048–54.

Campbell, Margaret. 2017. "Disabilities and Sexual Expression: A Review of the Literature." *Sociology Compass* 11 (9): e12508.

Canseco-Alba, Ana, and Gabriela Rodríguez-Manzo. 2019. "Sexual Interaction Is Essential for the Transformation of Non-Copulating Rats into Sexually Active Animals by the Endocannabinoid Anandamide." *Behavioural Brain Research* 359 (February): 418–27.

Carlson, Elof Axel. 1981. *Genes, Radiation, and Society: The Life and Work of H. J. Muller.* Cornell University Press.

Carr, Brenda. 1993. "'A Woman Speaks . . . I Am Woman and Not White': Politics of Voice, Tactical Essentialism, and Cultural Intervention in Audre Lorde's Activist Poetics and Practice." *College Literature* 20 (2): 133–53.

Carrigan, Mark. 2011. "There's More to Life Than Sex? Difference and Commonality Within the Asexual Community." *Sexualities* 14 (4): 462–78.

Carrigan, Mark, Kristina Gupta, and Todd Morrison, eds. 2014. *Asexuality and Sexual Normativity: An Anthology.* 1st ed. London New York: Routledge.

Carroll, Megan. 2024. "Asexuality." In *The Sage Encyclopedia of LGBTQ+ Studies,* edited by Abbie E. Goldberg, 115–19. 2nd ed. SAGE Publications.

Casey. 2018. "Pandas Are Asexual." *LGBT+* (blog). Amino, June 4. https://aminoapps .com/c/lgbt-1/page/blog/pandas-are-asexual/06wW_M4bSkuKJzbJroPZx Pwwnw8b1l0616l.

Castro, Joseph. 2016. "The Complicated Sex Lives of Giant Pandas." *Livescience.com,* September 27. https://www.livescience.com/56269-animal-sex-giant-pandas.html.

Catri, Florencia. 2021. "Defining Asexuality as a Sexual Identity: Lack/Little Sexual Attraction, Desire, Interest and Fantasies." *Sexuality & Culture* 25 (4): 1529–39.

Cera, Nicoletta, Ezio Domenico Di Pierro, Antonio Ferretti, Armando Tartaro, Gian Luca Romani, and Mauro Gianni Perrucci. 2014. "Brain Networks During Free Viewing of Complex Erotic Movie: New Insights on Psychogenic Erectile Dysfunction." *PLOS ONE* 9 (8): e105336.

Cerankowski, K. J. 2021. "The 'End' of Orgasm: The Erotics of Durational Pleasures." *Studies in Gender and Sexuality* 22 (3): 132–46.

Chan, Randolph C. H., and Janice Sin Yu Leung. 2023. "Experiences of Minority Stress and Their Impact on Suicidality Among Asexual Individuals." *Journal of Affective Disorders* 325 (March): 794–803.

Chao, Sophie, Karin Bolender, and Eben Kirksey, eds. 2022. *The Promise of Multispecies Justice.* Duke University Press.

Charlesworth, Brian. 2004. "John Maynard Smith." *Genetics* 168 (3): 1105–9.

Charnas, Suzy McKee. 1978. *Motherlines.* 1st ed. Putnam Publishing Group.

Chasin, CJ DeLuzio. 2013. "Reconsidering Asexuality and Its Radical Potential." *Feminist Studies* 39 (2): 405–26.

Chasnov, Jeffrey R. 2013. "The Evolutionary Role of Males in *C. elegans.*" *Worm* 2 (1): e21146.

Chatel, Amanda. 2017. "Why Pornhub Wants You to Have Sex in a Panda Costume Today." *Bustle,* March 16. https://www.bustle.com/p/why-pornhub-wants-you -to-have-sex-in-a-panda-costume-today-44412.

Chen, Lian, and John J. Wiens. 2021. "Multicellularity and Sex Helped Shape the Tree of Life." *Proceedings of the Royal Society B: Biological Sciences* 288 (1955): 20211265.

Chen, Mel Y. 2012. *Animacies: Biopolitics, Racial Mattering, and Queer Affect.* Duke University Press.

Chen, Pin-Hao A., Eshin Jolly, Jin Hyun Cheong, and Luke J. Chang. 2020. "Intersubject Representational Similarity Analysis Reveals Individual Variations in Affective Experience When Watching Erotic Movies." *NeuroImage* 216 (August): 116851.

Cheng, Joseph C., Joseph Secondary, William H. Burke, J. Paul Fedoroff, and R. Gregg Dwyer. 2015. "Neuroimaging and Sexual Behavior: Identification of Regional and Functional Differences." *Current Psychiatry Reports* 17 (7): 55.

Chenier, Elise. 2012. "The Natural Order of Disorder: Pedophilia, Stranger Danger and the Normalising Family." *Sexuality & Culture* 16 (2): 172–86.

Cho, Sumi, Kimberlé Williams Crenshaw, and Leslie McCall. 2013. "Toward a Field of Intersectionality Studies: Theory, Applications, and Praxis." *Signs* 38 (4): 785–810.

Ciaccio, Valentina, and Dina Di Giacomo. 2022. "Psychological Factors Related to Impotence as a Sexual Dysfunction in Young Men: A Literature Scan for Noteworthy Research Frameworks." *Clinics and Practice* 12 (4): 501–12.

Cikara, Mina, Jennifer L. Eberhardt, and Susan T. Fiske. 2010. "From Agents to Objects: Sexist Attitudes and Neural Responses to Sexualized Targets." *Journal of Cognitive Neuroscience* 23 (3): 540–51.

Cipolla, Cyd, and Kristina Gupta. 2017. "Neurogenderings and Neuroethics." In *The Routledge Handbook of Neuroethics*, edited by L. Syd M. Johnson and Karen S. Rommelfanger, 381–93. Routledge.

Clark, Mertice M., and Bennett G. Galef. 2000. "Why Some Male Mongolian Gerbils May Help at the Nest: Testosterone, Asexuality and Alloparenting." *Animal Behaviour* 59 (4): 801–6.

Clark, Mertice M., Leanne Tucker, and Bennett G. Galef Jr. 1992. "Stud Males and Dud Males: Intra-Uterine Position Effects on the Reproductive Success of Male Gerbils." *Animal Behaviour* 43 (2): 215–21.

Clarke, Esther, Katie Bradshaw, Kieran Drissell, Parag Kadam, Nikki Rutter, and Stefano Vaglio. 2022. "Primate Sex and Its Role in Pleasure, Dominance and Communication." *Animals: An Open Access Journal from MDPI* 12 (23): 3301.

Cohen-Kettenis, Peggy T., and Friedemann Pfäfflin. 2009. "The DSM Diagnostic Criteria for Gender Identity Disorder in Adolescents and Adults." *Archives of Sexual Behavior* 39 (2): 499–513.

Colegrave, Nick. 2012. "The Evolutionary Success of Sex." *EMBO Reports* 13 (9): 774–78.

Collard, Rosemary-Claire. 2013. "Panda Politics." *Canadian Geographer* 57 (2): 226–32.

Collins, Patricia Hill. 2005. "Black Women and Motherhood." In *Motherhood and Space: Configurations of the Maternal Through Politics, Home, and the Body*, edited by Sarah Hardy and Caroline Wiedmer, 149–59. Palgrave Macmillan US.

Conley, Terri D., Amy C. Moors, Jes L. Matsick, Ali Ziegler, and Brandon A. Valentine. 2011. "Women, Men, and the Bedroom Methodological and Conceptual Insights That Narrow, Reframe, and Eliminate Gender Differences in Sexuality." *Current Directions in Psychological Science* 20 (5): 296–300.

Conn, David Bruce, ed. 1991. *Atlas of Invertebrate Reproduction and Development*. 1st ed. Wiley.

Connell, R. W. 2005. *Masculinities*. 2nd ed. University of California Press.

Conrad, Peter, and Joseph W. Schneider. 1992. *Deviance and Medicalization: From Badness to Sickness*. Temple University Press.

Copulsky, Daniel, and Phillip L. Hammack. 2023. "Asexuality, Graysexuality, and Demisexuality: Distinctions in Desire, Behavior, and Identity." *Journal of Sex Research* 60 (2): 221–30.

Costa, Marco, Christoph Braun, and Niels Birbaumer. 2003. "Gender Differences in Response to Pictures of Nudes: A Magnetoencephalographic Study." *Biological Psychology* 63 (2): 129–47.

Cowan, Ruth Schwartz. 2016. "Commentary: Before Weismann and Germplasm There Was Galton and Eugenics: The Biological and Political Meaning of the Inheritance of Acquired Characteristics in the Late 19th Century." *International Journal of Epidemiology* 45 (1): 15–20.

Crow, James F. 1994. "Advantages of Sexual Reproduction." *Developmental Genetics* 15 (3): 205–13.

Crow, James F. 2005. "Hermann Joseph Muller, Evolutionist." *Nature Reviews Genetics* 6 (12): 941–45.

Cruzan, Mitchell B. 2018. *Evolutionary Biology: A Plant Perspective.* Oxford University Press.

Cryle, Peter. 2008. "Building a Sexological Concept Through Fictional Narrative." *French Cultural Studies* 19 (2): 115–40.

Cryle, Peter, and Alison Moore. 2012. *Frigidity: An Intellectual History.* Palgrave Macmillan.

Curtis, Debra. 2004. "Commodities and Sexual Subjectivities: A Look at Capitalism and Its Desires." *Cultural Anthropology* 19 (1): 95–121.

Cuthbert, Karen. 2015. "You Have to Be Normal to Be Abnormal: An Empirically Grounded Exploration of the Intersection of Asexuality and Disability." *Sociology* 51 (2): 241–57.

Cuthbert, Karen. 2022. "Asexuality and Epistemic Injustice: A Gendered Perspective." *Journal of Gender Studies* 31 (7): 840–51.

Dalziel, Anne C., Svetlana Tirbhowan, Hayley F. Drapeau, et al. 2020. "Using Asexual Vertebrates to Study Genome Evolution and Animal Physiology: Banded (*Fundulus diaphanus*) × Common Killifish (*F. heteroclitus*) Hybrid Lineages as a Model System." *Evolutionary Applications* 13 (6): 1214–39.

Dana, Geraldina. 2020. "La comunidad virtual de asexuales del área metropolitana de Buenos Aires." *Sexualidad, salud y sociedad (Rio de Janeiro)* 34 (April): 126–52.

Dang, Silvain, Sabrina Chang, and Lori A. Brotto. 2017. "The Lived Experiences of Sexual Desire Among Chinese-Canadian Men and Women." *Journal of Sex & Marital Therapy* 43 (4): 306–25.

Dapper, Amy L., and Bret A. Payseur. 2017. "Connecting Theory and Data to Understand Recombination Rate Evolution." *Philosophical Transactions of the Royal Society B: Biological Sciences* 372 (1736): 20160469.

Davidson, Arnold. 2001. *The Emergence of Sexuality: Historical Epistemology and the Formation of Concepts.* Harvard University Press.

Dawson, Matt, Liz McDonnell, and Susie Scott. 2016. "Negotiating the Boundaries of Intimacy: The Personal Lives of Asexual People." *Sociological Review* 64 (2): 349–65.

Deacon, Susan, Victor Minichiello, and David Plummer. 1995. "Sexuality and Older People: Revisiting the Assumptions." *Educational Gerontology* 21 (5): 497–513.

Decker, Julie Sondra. 2015. *The Invisible Orientation: An Introduction to Asexuality.* Reprint ed. Skyhorse.

Deen, Aminu. 2008. "Testosterone Profiles and Their Correlation with Sexual Libido in Male Camels." *Research in Veterinary Science* 85 (2): 220–26.

"Definition of EROTIC." n.d. Accessed September 25, 2023. https://www.merriam
-webster.com/dictionary/erotic.

Dell'Amore, Christine. 2013. "Is Breeding Pandas in Captivity Worth It?" *National
Geographic*, August 27. https://www.nationalgeographic.com/science/article
/130827-giant-panda-national-zoo-baby-breeding-animals-science.

de Meeûs, Thierry, Franck Prugnolle, and Philip Agnew. 2007. "Asexual Reproduction:
Genetics and Evolutionary Aspects." *Cellular and Molecular Life Sciences* 64 (11):
1355–72.

Dennis, Rutledge M. 1995. "Social Darwinism, Scientific Racism, and the
Metaphysics of Race." *Journal of Negro Education* 64 (3): 243–52.

Dhillon, Sohita, and Susan J. Keam. 2019. "Bremelanotide: First Approval." *Drugs* 79
(14): 1599–1606.

Diamond, Lisa, and Janna Dickenson. 2012. "The Neuroimaging of Love and Desire:
Review and Future Directions." *Clinical Neuropsychiatry* 9 (1): 39–46.

Diamond, Shaindl. 2017. "Trapped in Change: Using Queer Theory to Examine
the Progress of Psy-Theories and Interventions with Sexuality and Gender."
In *Routledge International Handbook of Critical Mental Health*, edited by Bruce
Cohen, 89–97. Routledge.

Di Chiro, Giovanna. 2010. "Polluted Politics? Confronting Toxic Discourse, Sex Panic,
and Eco-Normativity." In *Queer Ecologies: Sex, Nature, Politics, Desire*, by Catriona
Mortimer-Sandilands and Bruce Erickson, 199–230. Indiana University Press.

Dixson, Alan F. 2012. *Primate Sexuality Comparative Studies of the Prosimians,
Monkeys, Apes and Human Beings*. 2nd ed. Oxford University Press.

Dockrill, Peter. 2019. "This Wild New Hypothesis Says Sex May Have Evolved to
Fight Certain Types of Cancer." *ScienceAlert* (blog), June 6. https://www
.sciencealert.com/sex-might-have-evolved-to-fight-cancers-that-can-be
-transmitted-like-viruses.

Doums, Claudie. 2021. "Parthenogenesis." In *Encyclopedia of Social Insects*, edited by
Christopher K. Starr, 723–28. Springer.

Drescher, Jack. 2015. "Out of DSM: Depathologizing Homosexuality." *Behavioral
Sciences* 5 (4): 565–75.

Drucker, Peter. 2015. *Warped: Gay Normality and Queer Anti-Capitalism*. Brill.

Drury, Jonathan P. 2013. "Sex, Gender, and Evolution Beyond Genes." In *Challenging
Popular Myths of Sex, Gender and Biology*, edited by Malin Ah-King, 43–52.
Springer International Publishing.

Duchesne, Annie, and Anelis Kaiser Trujillo. 2021. "Reflections on Neurofeminism
and Intersectionality Using Insights from Psychology." *Frontiers in Human
Neuroscience* 15 (September 28): 684412.

Duncombe, Jean, and Dennis Marsden. 1996. "Whose Orgasm Is This Anyway? Sex
Work in Long-Term Couple Relationships." In *Sexual Cultures*, edited by Jeffrey
Weeks and Janet Holland, 220–38. Macmillan.

Dussauge, Isabelle. 2013. "The Experimental Neuro-Framing of Sexuality." *Graduate
Journal of Social Science* 10 (1): 124–51.

Dussauge, Isabelle, and Anelis Kaiser. 2012. "Re-Queering the Brain." In *Neurofeminism: Issues at the Intersection of Feminist Theory and Cognitive Science*, edited by Robyn Bluhm, Anne Jaap Jacobson, and Heidi Lene Maibom, 121–44. Palgrave Macmillan UK.

Edelman, Lee. 2004. *No Future: Queer Theory and the Death Drive*. Duke University Press.

Egan, Kristen R. 2011. "Conservation and Cleanliness: Racial and Environmental Purity in Ellen Richards and Charlotte Perkins Gilman." *WSQ: Women's Studies Quarterly* 39 (3): 77–92.

Elliott, Mark L. 1985. "The Use of 'Impotence' and 'Frigidity': Why Has 'Impotence' Survived." *Journal of Sex & Marital Therapy* 11 (1): 51–56.

Ellis, Havelock. 1936. *Studies in the Psychology of Sex*. Vol. 1. Random House.

Emens, Elizabeth F. 2013. "Compulsory Sexuality." *Stanford Law Review* 66 (March): 303–86.

Eng, David L. 2001. *Racial Castration: Managing Masculinity in Asian America*. Duke University Press.

Engelstädter, Jan. 2008. "Constraints on the Evolution of Asexual Reproduction." *BioEssays* 30 (11–12): 1138–50.

Enszer, Julie R. 2016. "'How to Stop Choking to Death': Rethinking Lesbian Separatism as a Vibrant Political Theory and Feminist Practice." *Journal of Lesbian Studies* 20 (2): 180–96.

Erickson, Mark. 2016. *Science, Culture and Society: Understanding Science in the 21st Century*. 2nd ed. Polity.

Erickson-Schroth, Laura. 2013. "Update on the Biology of Transgender Identity." *Journal of Gay & Lesbian Mental Health* 17 (2): 150–74.

Esmail, Shaniff, Kim Darry, Ashlea Walter, and Heidi Knupp. 2010. "Attitudes and Perceptions Towards Disability and Sexuality." *Disability and Rehabilitation* 32 (14): 1148–55.

Ethofer, Thomas, Sarah Wiethoff, Silke Anders, Benjamin Kreifelts, Wolfgang Grodd, and Dirk Wildgruber. 2007. "The Voices of Seduction: Cross-Gender Effects in Processing of Erotic Prosody." *Social Cognitive and Affective Neuroscience* 2 (4): 334–37.

Falk, Emily B., Luke W. Hyde, Colter Mitchell, et al. 2013. "What Is a Representative Brain? Neuroscience Meets Population Science." *Proceedings of the National Academy of Sciences* 110 (44): 17615–22.

Farah, Martha J. 2014. "Brain Images, Babies, and Bathwater: Critiquing Critiques of Functional Neuroimaging." *The Hastings Center Report* Spec No (April): S19–30.

Fausto-Sterling, Anne. 2000. "Beyond Difference: Feminism and Evolutionary Psychology." In *Alas, Poor Darwin: Arguments against Evolutionary Psychology*, edited by Hilary Rose and Steven Rose, 209–27. Harmony Books.

Fausto-Sterling, Anne. 2008. *Myths of Gender: Biological Theories About Women And Men*. Revised ed. Basic Books.

Feng, Chunliang, Lili Wang, Naiyi Wang, Ruolei Gu, and Yue-Jia Luo. 2012. "The Time Course of Implicit Processing of Erotic Pictures: An Event-Related Potential Study." *Brain Research* 1489 (December): 48–55.

Fernandez, Colin. 2016. "A Passion for Survival: How Reproducing Sexually Makes Us Much More Resistant to Infections." *Mail Online*, December 21, sec. Science. http://www.dailymail.co.uk/~/article-4053760/index.html.

Ferretti, Antonio, Massimo Caulo, Cosimo Del Gratta, et al. 2005. "Dynamics of Male Sexual Arousal: Distinct Components of Brain Activation Revealed by fMRI." *Neuroimage* 26 (4): 1086–96.

Fisher, Helen, Arthur Aron, and Lucy L. Brown. 2005. "Romantic Love: An fMRI Study of a Neural Mechanism for Mate Choice." *Journal of Comparative Neurology* 493 (1): 58–62.

Flore, Jacinthe. 2020. *A Genealogy of Appetite in the Sexual Sciences.* Springer International Publishing.

Fontaneto, Diego, and Timothy G. Barraclough. 2015. "Do Species Exist in Asexuals? Theory and Evidence from Bdelloid Rotifers." *Integrative and Comparative Biology* 55 (2): 253–63.

Fonteille, Véronique, and Serge Stoléru. 2011. "The Cerebral Correlates of Sexual Desire: Functional Neuroimaging Approach." *Sexologies* 20 (3): 142–48.

Fooladi, Ensieh, Rakibul M. Islam, Robin J. Bell, Penelope J. Robinson, Maryam Masoumi, and Susan R. Davis. 2020. "The Prevalence of Hypoactive Sexual Desire Disorder in Australian and Iranian Women at Midlife." *Menopause* 27 (11): 1274.

Foster, Aasha B., Austin Eklund, Melanie E. Brewster, Amelia D. Walker, and Emma Candon. 2019. "Personal Agency Disavowed: Identity Construction in Asexual Women of Color." *Psychology of Sexual Orientation and Gender Diversity* 6 (2): 127–37.

Foucault, Michel. 1990. *The History of Sexuality.* Vol. 1, *An Introduction.* Translated by Robert Hurley. Vintage Books, Random House.

Fradin, Hélène, Karin Kiontke, Charles Zegar, et al. 2017. "Genome Architecture and Evolution of a Unichromosomal Asexual Nematode." *Current Biology* 27 (19): 2928–2939.e6.

French, Kara M. 2021. *Against Sex: Identities of Sexual Restraint in Early America.* University of North Carolina Press.

Freud, Sigmund. (1915) 1962. "Instincts and Their Vicissitudes." In *The Standard Edition of the Complete Psychological Works of Sigmund Freud*, edited and translated by James Strachey, vol. 14, 109–40. Hogarth Press.

Freud, Sigmund. 1964. "Femininity." In *New Introductory Lectures on Psycho-Analysis: The Standard Edition*, by Sigmund Freud, 139–67. W. W. Norton & Company.

Freud, Sigmund. 2000. *Three Essays on the Theory of Sexuality.* Revised ed. Basic Books.

Friedrichs, Kassandra, and Philipp Kellmeyer. 2022. "Neurofeminism: Feminist Critiques of Research on Sex/Gender Differences in the Neurosciences." *European Journal of Neuroscience* 56 (11): 5987–6002.

Frigerio, Alberto, Lucia Ballerini, and Maria Valdés Hernández. 2021. "Structural, Functional, and Metabolic Brain Differences as a Function of Gender Identity or Sexual Orientation: A Systematic Review of the Human Neuroimaging Literature." *Archives of Sexual Behavior* 50 (8): 3329–52.

Frye, Marilyn. 1997. "Some Reflections on Separatism and Power." In *Feminist Social Thought*, edited by Diana Tietjens Meyers, 406–14. Routledge.

Fuechtner, Veronika, Douglas E. Haynes, and Ryan M. Jones. 2018. *A Global History of Sexual Science, 1880–1960*. University of California Press.

Furuichi, Takeshi, Richard Connor, and Chie Hashimoto. 2014. "Non-Conceptive Sexual Interactions in Monkeys, Apes, and Dolphins." In *Primates and Cetaceans: Field Research and Conservation of Complex Mammalian Societies*, edited by Juichi Yamagiwa and Leszek Karczmarski, 385–408. Primatology Monographs. Springer Japan.

Futuyma, Douglas J., and Stephen C. Stearns. 2010. "George Christopher Williams, 1926–2010." *Evolution* 64 (12): 3339–43.

Gaillard, Alexandra, Daniel J. Fehring, and Susan L. Rossell. 2021. "Sex Differences in Executive Control: A Systematic Review of Functional Neuroimaging Studies." *European Journal of Neuroscience* 53 (8): 2592–2611.

Galis, Frietson, and Jacques J. M. van Alphen. 2020. "Parthenogenesis and Developmental Constraints." *Evolution & Development* 22 (1–2): 205–17.

Ganesan, Asha, James S. Morandini, Aaron Veldre, Kevin J. Hsu, and Ilan Dar-Nimrod. 2020. "Ethnic Differences in Visual Attention to Sexual Stimuli Among Asian and White Heterosexual Women and Men." *Personality and Individual Differences* 155 (March): 109630.

Gannon, Megan. 2015. "San Diego Zoo Turns Off Panda Cam for Mating Time." Livescience.com, March 11. https://www.livescience.com/50109-san-diego-zoo-pandas-mating-season.html.

Garland-Thomson, Rosemarie. 2011. "Misfits: A Feminist Materialist Disability Concept." *Hypatia* 26 (3): 591–609.

Garland-Thomson, Rosemarie. 2012. "The Case for Conserving Disability." *Journal of Bioethical Inquiry* 9 (3): 339–55.

Geddes, Linda. 2011. "Sex on the Brain: What Turns Women On, Mapped Out." *New Scientist*, August 5. https://www.newscientist.com/article/dn20770-sex-on-the-brain-what-turns-women-on-mapped-out/.

Georgiadis, Janniko R. 2011. "Exposing Orgasm in the Brain: A Critical Eye." *Sexual and Relationship Therapy* 26 (4): 342–55.

Georgiadis, Janniko R., and Morten L. Kringelbach. 2012. "The Human Sexual Response Cycle: Brain Imaging Evidence Linking Sex to Other Pleasures." *Progress in Neurobiology* 98 (1): 49–81.

Georgiadis, Janniko R., Morten L. Kringelbach, and James G. Pfaus. 2012. "Sex for Fun: A Synthesis of Human and Animal Neurobiology." *Nature Reviews Urology* 9 (9): 486–99.

Gerhard, Jane. 2000. "Revisiting 'The Myth of the Vaginal Orgasm': The Female Orgasm in American Sexual Thought and Second Wave Feminism." *Feminist Studies* 26 (2): 449–76.

Getzlaff, J. A. 2000. "Giant Pandas to Be Given Viagra." *Salon*, April 26. https://www .salon.com/2000/04/26/pandas/.

Gherovici, Patricia. 2011. "Psychoanalysis Needs a Sex Change." *Gay and Lesbian Issues and Psychology Review* 7 (January): 3–18.

Gibson, Amanda K., Lynda F. Delph, and Curtis M. Lively. 2017. "The Two-Fold Cost of Sex: Experimental Evidence from a Natural System." *Evolution Letters* 1 (1): 6–15.

Gilarek, Anna. 2013. "Different Feminist Approaches to Reproductive Technologies: Biopolitics in Feminist Speculative Fiction." In *Esthetic Experiments: Interdisciplinary Challenges in American Studies*, edited by Edyta Just and Marek M. Wojtaszek, 73–88. Cambridge Scholars Publishing.

Gillath, Omri, and Melanie Canterberry. 2012. "Neural Correlates of Exposure to Subliminal and Supraliminal Sexual Cues." *Social Cognitive and Affective Neuroscience* 7 (8): 924–36.

Gillis-Buck, Eva Mae. 2016. "Redefining 'Virgin Birth' After Kaguya: Mammalian Parthenogenesis in Experimental Biology, 2004–2014." *Catalyst: Feminism, Theory, Technoscience* 2 (1): 1–68.

Gill-Peterson, Jules. 2018. *Histories of the Transgender Child*. University of Minnesota Press.

Gilman, Charlotte Perkins. (1915) 1979. *Herland*. Knopf Doubleday Publishing Group.

Ginzberg, Ruth. 1992. "Audre Lorde's (Nonessentialist) Lesbian Eros." *Hypatia* 7 (4): 73–90.

Gizewski, Elke R., Eva Krause, Marc Schlamann, et al. 2009. "Specific Cerebral Activation Due to Visual Erotic Stimuli in Male-to-Female Transsexuals Compared with Male and Female Controls: An fMRI Study." *Journal of Sexual Medicine* 6 (2): 440–48.

Glémin, Sylvain, Clémentine M. François, and Nicolas Galtier. 2019. "Genome Evolution in Outcrossing vs. Selfing vs. Asexual Species." In *Evolutionary Genomics*, edited by Maria Anisimova, 331–69. Humana.

Glick, Elisa. 2000. "Sex Positive: Feminism, Queer Theory, and the Politics of Transgression." *Feminist Review*, no. 64 (April): 19–45.

Gochfeld, Michael, and Joanna Burger. 2011. "Disproportionate Exposures in Environmental Justice and Other Populations: The Importance of Outliers." *American Journal of Public Health* 101 (Supplement 1): S53–63.

Gola, Mateusz, Małgorzata Wordecha, Artur Marchewka, and Guillaume Sescousse. 2016. "Visual Sexual Stimuli—Cue or Reward? A Perspective for Interpreting Brain Imaging Findings on Human Sexual Behaviors." *Frontiers in Human Neuroscience* 10 (August 14). https://doi.org/10.3389/fnhum.2016.00402.

Gola, Mateusz, Małgorzata Wordecha, Guillaume Sescousse, et al. 2017. "Can Pornography Be Addictive? An fMRI Study of Men Seeking Treatment for Problematic Pornography Use." *Neuropsychopharmacology* 42 (10): 2021–31.

Goldhammer, Denisa, and Marita McCabe. 2011. "A Qualitative Exploration of the Meaning and Experience of Sexual Desire Among Partnered Women." *Canadian Journal of Human Sexuality* 20 (1–2): 19–29.

Goodenough, Ursula, and Joseph Heitman. 2014. "Origins of Eukaryotic Sexual Reproduction." *Cold Spring Harbor Perspectives in Biology* 6 (3): a016154.

Goodrich, Marcia. 2012. "Sex: It's a Good Thing, Evolutionarily Speaking." *Michigan Tech News* (blog), May 29. https://www.mtu.edu/news/2012/05/sex-its-good -thing-evolutionarily-speaking.html.

Gooren, Louis. 2006. "The Biology of Human Psychosexual Differentiation." *Hormones and Behavior* 50 (4): 589–601.

Graham, Laurie A., and Peter L. Davies. 2021. "Horizontal Gene Transfer in Vertebrates: A Fishy Tale." *Trends in Genetics* 37 (6): 501–3.

Grandoni, Dino. 2023. "This Snake Can Make Babies Without a Mate." *Washington Post*, February 14, sec. Environment.

Grasswick, Heidi. 2021. "Feminist Epistemology." In *The Oxford Handbook of Feminist Philosophy*, edited by Kim Q. Hall and Ásta, 198–212. Oxford University Press.

Gray, Michael W. 2017. "Lynn Margulis and the Endosymbiont Hypothesis: 50 Years Later." *Molecular Biology of the Cell* 28 (10): 1285–87.

Gray, Russell 1997. "'In the Belly of the Monster': Feminism, Developmental Systems, and Evolutionary Explanations." In *Feminism and Evolutionary Biology*, edited by Patricia Adair Gowaty, 385–413. Springer.

Griffiths, David. 2015. "Queer Theory for Lichens." *UnderCurrents: Journal of Critical Environmental Studies* 19 (October): 36–45.

Groneman, Carol. 1994. "Nymphomania: The Historical Construction of Female Sexuality." *Signs* 19 (2): 337–67.

Groneman, Carol. 2001. *Nymphomania: A History*. W. W. Norton & Company.

Grosz, Elizabeth. 2011. *Becoming Undone: Darwinian Reflections on Life, Politics, and Art*. Duke University Press.

Guay, Andre T. 2005. "Commentary on Androgen Deficiency in Women and the FDA Advisory Board's Recent Decision to Request More Safety Data." *International Journal of Impotence Research* 17 (4): 375–76.

Gupta, Kristina. 2012a. "Sex (in the) Machine." *Neuroethics Blog*, May 31. http://www .theneuroethicsblog.com/2012/05/sex-in-machine.html.

Gupta, Kristina. 2012b. "Why Do Voles Fall in Love? Interview with Feminist Science Studies Scholar Angela Willey." *Neuroethics Blog*, June 28. http://www .theneuroethicsblog.com/2012/06/why-do-voles-fall-in-love-interview.html.

Gupta, Kristina. 2013a. "Happy Asexual Meets DSM." *SocialText Online (Periscope)*, October 24. https://socialtextjournal.org/periscope_article/happy-asexual -meets-dsm/.

Gupta, Kristina. 2013b. "Picturing Space for Lesbian Nonsexualities: Rethinking Sex-Normative Commitments Through *The Kids Are All Right* (2010)." *Journal of Lesbian Studies* 17 (1): 103–18.

Gupta, Kristina. 2014. "Asexuality and Disability: Mutual Negation in *Adams v. Rice* and New Directions for Coalition Building." In *Asexualities: Feminist and Queer Perspectives*, edited by K. J. Cerankowski and Megan Milks, 238–301. Routledge.

Gupta, Kristina. 2015. "Compulsory Sexuality: Evaluating an Emerging Concept." *Signs: Journal of Women in Culture & Society* 41 (1): 131–54.

Gupta, Kristina. 2016. "What Does Asexuality Teach Us About Sexual Disinterest? Recommendations for Health Professionals Based on a Qualitative Study with Asexually Identified People." *Journal of Sex & Marital Therapy* 43 (1): 1–14.

Gupta, Kristina. 2017. "'And Now I'm Just Different, but There's Nothing Actually Wrong with Me': Asexual Marginalization and Resistance." *Journal of Homosexuality* 64 (8): 991–1013.

Gupta, Kristina. 2019a. "Gendering Asexuality and Asexualizing Gender: A Qualitative Study Exploring the Intersections Between Gender and Asexuality." *Sexualities* 22 (7–8): 1197–1216.

Gupta, Kristina. 2019b. *Medical Entanglements: Rethinking Feminist Debates About Healthcare*. Rutgers University Press.

Gupta, Kristina. 2022. "What Is a Sexual Act?" In *The Routledge Handbook of Philosophy of Sex and Sexuality*, edited by Brian D. Earp, Clare Chambers, and Lori Watson, 9–19. Routledge.

Halberg, Claus. 2022. "Neurosexism, Neurofeminism, and Neurocentrism: From Gendered Brains to Embodied Minds." *Nordic Journal of Feminist and Gender Research* 31 (3): 279–91.

Hall, Marny. 2004. "Resolving the Curious Paradox of the (A)Sexual Lesbian." *Journal of Couple & Relationship Therapy* 3 (2): 75–83.

Halley, Janet E. 1994. "Sexual Orientation and the Politics of Biology: A Critique of the Argument from Immutability." *Stanford Law Review* 46 (3): 503–68.

Hamann, Stephan, Rebecca A. Herman, Carla L. Nolan, and Kim Wallen. 2004. "Men and Women Differ in Amygdala Response to Visual Sexual Stimuli." *Nature Neuroscience* 7 (4): 411–16.

Hamann, Stephan, Jennifer Stevens, Janice Hassett Vick, et al. 2014. "Brain Responses to Sexual Images in 46,XY Women with Complete Androgen Insensitivity Syndrome Are Female-Typical." *Hormones and Behavior* 66 (5): 724–30.

Hamburger, Christian. 1953. "The Desire for Change of Sex as Shown by Personal Letters from 465 Men and Women." *Acta Endocrinologica* 14 (4): 361–75.

Hammonds, Evelynn M. 1994. "Black (W)holes and the Geometry of Black Female Sexuality." *Differences: A Journal of Feminist Cultural Studies* 6 (2–3): 126–46.

Hammonds, Evelynn M. 1996. "Toward a Genealogy of Black Female Sexuality: The Problematic of Silence." In *Feminist Genealogies, Colonial Legacies, Democratic Futures*, edited by M. Jacqui Alexander and Chandra Talpade Mohanty, 93–104. Routledge.

Han, Chong-suk. 2006. "Geisha of a Different Kind: Gay Asian Men and the Gendering of Sexual Identity." *Sexuality and Culture* 10 (3): 3–28.

Han, Maythe Seung-Won. 2022. "More-than-Human Kinship Against Proximal Loneliness: Practising Emergent Multispecies Care with a Dog in a Pandemic and Beyond." *Feminist Theory* 23 (1): 109–24.

Hancock, Ange-Marie. 2016. *Intersectionality: An Intellectual History*. Oxford University Press.

Hand, Melanie L., and Anna M. G. Koltunow. 2014. "The Genetic Control of Apomixis: Asexual Seed Formation." *Genetics* 197 (2): 441–50.

Hanson, Elizabeth Hanna. 2014. "Toward an Asexual Narrative Structure." In *Asexualities: Feminist and Queer Perspectives*, edited by K. J. Cerankowski and Megan Milks, 344–74. Routledge.

Haraway, Donna J. 1988. "Situated Knowledges: The Science Question in Feminism and the Privilege of Partial Perspective." *Feminist Studies* 14 (3): 575–99.

Haraway, Donna J. 1990. *Primate Visions: Gender, Race, and Nature in the World of Modern Science*. Reprint ed. Routledge.

Haraway, Donna J. 2003. *The Companion Species Manifesto: Dogs, People, and Significant Otherness*. Edited by Matthew Begelke. Prickly Paradigm Press.

Haraway, Donna J. 2008. *When Species Meet*. University of Minnesota Press.

Harding, Cheryl F., and Harvey H. Feder. 1976. "Relation Between Individual Differences in Sexual Behavior and Plasma Testosterone Levels in the Guinea Pig." *Endocrinology* 98 (5): 1198–1205.

Harding, Sandra. 1998. *Is Science Multicultural? Postcolonialisms, Feminisms, and Epistemologies*. Indiana University Press.

Hargons, Candice Nicole, Jardin Dogan, Natalie Malone, Shemeka Thorpe, Della V. Mosley, and Danelle Stevens-Watkins. 2021. "Balancing the Sexology Scales: A Content Analysis of Black Women's Sexuality Research." *Culture, Health & Sexuality* 23 (9): 1287–1301.

Harshaw, Christopher. 2008. "Alimentary Epigenetics: A Developmental Psychobiological Systems View of the Perception of Hunger, Thirst and Satiety." *Developmental Review: DR* 28 (4): 541–69.

Hartfield, Matthew, and Peter D. Keightley. 2012. "Current Hypotheses for the Evolution of Sex and Recombination." *Integrative Zoology* 7 (2): 192–209.

Hartley, Heather. 2006. "The 'Pinking' of Viagra Culture: Drug Industry Efforts to Create and Repackage Sex Drugs for Women." *Sexualities* 9 (3): 363–78.

Hedrick, Philip W. 1987. "Population Genetics of Intragametophytic Selfing." *Evolution* 41 (1): 137–44.

Heijkoop, Roy, Patty T. Huijgens, and Eelke M. S. Snoeren. 2018. "Assessment of Sexual Behavior in Rats: The Potentials and Pitfalls." *Behavioural Brain Research*, Animal Model of the Year 2036: Novel Perspectives in Behavioral Neuroscience, 352 (October): 70–80.

Heinzel, Alexander, Martin Walter, Felix Schneider, et al. 2006. "Self-Related Processing in the Sexual Domain: A Parametric Event-Related fMRI Study Reveals

Neural Activity in Ventral Cortical Midline Structures." *Social Neuroscience* 1 (1): 41–51.

Held, Mirjam B. E. 2023. "Decolonizing Science: Undoing the Colonial and Racist Hegemony of Western Science." *Journal of MultiDisciplinary Evaluation* 19 (44): 88–101.

Helm, Bennett. 2009. "Friendship." In *The Stanford Encyclopedia of Philosophy*, edited by Edward N. Zalta. http://plato.stanford.edu/archives/fall2009/entries /friendship/.

Hennessy, Rosemary. 2000. *Profit and Pleasure: Sexual Identities in Late Capitalism*. Routledge.

Henrich, Joseph, Steven J. Heine, and Ara Norenzayan. 2010. "The Weirdest People in the World?" *Behavioral and Brain Sciences* 33 (2–3): 61–83.

Hermann, Lea, Ai Baba, D. Montagner, et al. 2022. "2020 Ace Community Survey Summary Report." The Ace Community Survey Team.

Hiebert, Laurel S., Carl Simpson, and Stefano Tiozzo. 2021. "Coloniality, Clonality, and Modularity in Animals: The Elephant in the Room." *Journal of Experimental Zoology Part B: Molecular and Developmental Evolution* 336 (3): 198–211.

Higginbotham, Evelyn Brooks. 1992. "African-American Women's History and the Metalanguage of Race." *Signs: Journal of Women in Culture and Society* 17 (2): 251–74.

Higginbotham, Evelyn Brooks. 1994. *Righteous Discontent: The Women's Movement in the Black Baptist Church, 1880–1920*. Harvard University Press.

Hill, Darryl B. 2005. "Sexuality and Gender in Hirschfeld's *Die Transvestiten*: A Case of the 'Elusive Evidence of the Ordinary.'" *Journal of the History of Sexuality* 14 (3): 316–32.

Hill, Geoffrey E. 2019. *Mitonuclear Ecology*. Oxford Series in Ecology and Evolution. Oxford University Press.

Hinderliter, Andrew C. 2009. "Asexuality: The History of a Definition." *Asexual Explorations* (blog). https://www.asexualexplorations.net/home/history_of _definition.html.

Hinderliter, Andrew. 2015. "Sexual Dysfunctions and Asexuality in DSM-5." In *The DSM-5 in Perspective: Philosophical Reflections on the Psychiatric Babel*, edited by Steeves Demazeux and Patrick Singy, 125–39. Springer.

Hine, Darlene Clark. 1989. "Rape and the Inner Lives of Black Women in the Middle West." *Signs: Journal of Women in Culture and Society* 14 (4): 912–20.

Hine, Robert. 2019. "Sexual Intercourse." In *A Dictionary of Biology*, edited by Robert Hine. Oxford University Press.

Hintze, Sara, Desiree Scott, Simon Turner, Simone L. Meddle, and Richard B. D'Eath. 2013. "Mounting Behaviour in Finishing Pigs: Stable Individual Differences Are Not Due to Dominance or Stage of Sexual Development." *Applied Animal Behaviour Science* 147 (1): 69–80.

Hird, Myra J. 2006. "Animal Transex." *Australian Feminist Studies* 21 (49): 35–50.

Hird, Myra J. 2009a. "Feminist Engagements with Matter." *Feminist Studies* 35 (2): 329–47.

Hird, Myra J. 2009b. *The Origins of Sociable Life: Evolution After Science Studies.* Palgrave Macmillan UK.

Hird, Myra J. 2012. "Digesting Difference: Metabolism and the Question of Sexual Difference." *Configurations* 20 (3): 213–37.

Hirschfeld, Magnus. 1910. *Die Transvestiten: Eine Untersuchung über den erotischen Verkleidungstrieb: Mit umfangreichen casuistischen und historischen Material.* Medicinischer Verlag.

Hirschfeld, Magnus. 1918. *Sexualpathologie, v.1, 1918.* A. Marcus & E. Webers.

Hirschfeld, Magnus. 1921. *Sexualpathologie: Ein Lehrbuch für Aerzte und Studierende.* Marcus & Weber. http://archive.org/details/b20442245.

Hirschfeld, Magnus. 1932. *Sexual Pathology: Being a Study of the Abnormalities of the Sexual Functions.* Translated by Jerome Gibbs. Julian Press.

Hirschfeld, Magnus. 1991. *Transvestites: The Erotic Drive to Cross-Dress.* Prometheus Books.

historicallyace. 2016. "What Kind of Attraction? A History of the Split Attraction Model." *Tumblr* (blog), October 24. https://historicallyace.tumblr.com/post /152267147477/what-kind-of-attraction-a-history-of-the-split.

Hochwagen, Andreas. 2008. "Meiosis." *Current Biology* 18 (15): R641–45.

Hofstatter, Paulo G., and Daniel J. G. Lahr. 2019. "All Eukaryotes Are Sexual, Unless Proven Otherwise: Many So-Called Asexuals Present Meiotic Machinery and Might Be Able to Have Sex." *BioEssays: News and Reviews in Molecular, Cellular and Developmental Biology* 41 (6): e1800246.

Hojsgaard, Diego, and Manfred Schartl. 2021. "Skipping Sex: A Nonrecombinant Genomic Assemblage of Complementary Reproductive Modules." *BioEssays* 43 (1): 2000111.

Horgan, John. 2011. "R. I. P. Lynn Margulis, Biological Rebel." Scientific American Blog Network, November 24. https://blogs.scientificamerican.com/cross -check/r-i-p-lynn-margulis-biological-rebel/.

Horta, Oscar, and Frauke Albersmeier. 2020. "Defining Speciesism." *Philosophy Compass* 15 (11): e12708.

Horton-Stallings, LaMonda. 2007. *Mutha' Is Half a Word: Intersections of Folklore, Vernacular, Myth, and Queerness in Black Female Culture.* Ohio State University Press.

Houar, Maria Teresa. 2022. "The Erotic of Abstinence: Refusing the White-Possessive and Embracing Settler Abstinence in Performance Pedagogy." In *Stages of Reckoning*, edited by Amy Mihyang Ginther, 95–111. Routledge.

Houdenhove, Ellen Van, Luk Gijs, Guy T'Sjoen, and Paul Enzlin. 2015. "Stories About Asexuality: A Qualitative Study on Asexual Women." *Journal of Sex & Marital Therapy* 41 (3): 262–81.

Hu, S. H., N. Wei, Q.-D. Wang, et al. 2008. "Patterns of Brain Activation During Visually Evoked Sexual Arousal Differ Between Homosexual and Heterosexual Men." *American Journal of Neuroradiology* 29 (10): 1890–96.

Hubbard, Ruth. 1990. *The Politics of Women's Biology*. Rutgers University Press.

Iasenza, Suzanne. 2002. "Beyond 'Lesbian Bed Death': The Passion and Play in Lesbian Relationships." *Journal of Lesbian Studies* 6 (1): 111.

Ibine, Bolade, Linda Sefakor Ametepe, Maxfield Okere, and Martina Anto-Ocrah. 2020. "'I Did Not Know It Was a Medical Condition': Predictors, Severity and Help Seeking Behaviors of Women with Female Sexual Dysfunction in the Volta Region of Ghana." *PLOS ONE* 15 (1): e0226404.

Iha, Higor A., Naofumi Kunisawa, Kentaro Tokudome, et al. 2017. "Immunohistochemical Analysis of Fos Protein Expression for Exploring Brain Regions Related to Central Nervous System Disorders and Drug Actions." In *In Vivo Neuropharmacology and Neurophysiology*, edited by Athineos Philippu, 389–408. Springer.

Irvine, Janice M. 1993. "Regulated Passions: The Invention of Inhibited Sexual Desire and Sex Addiction." *Social Text*, no. 37 (Winter): 203–26.

Irvine, Janice M. 2005. *Disorders of Desire: Sexuality and Gender in Modern American Sexology*. Revised and expanded ed. Temple University Press.

Jabr, Ferris. 2012. "In the Heat for a Moment: The Male Giant Panda's Sex Drive Fluctuates to Match the Female's Short-Lived Libido." *Scientific American*, April 4. https://www.scientificamerican.com/article/male-panda-sex-drive/.

Jackson, John P., and Nadine M. Weidman. 2004. *Race, Racism, and Science: Social Impact and Interaction*. Bloomsbury Academic.

Jackson, Stevi. 1996. "Heterosexuality as a Problem for Feminist Theory." In *Sexualizing the Social: Power and the Organization of Sexuality*, edited by Lisa Adkins and Vicki Merchant, 15–34. Palgrave Macmillan UK.

Jagose, Annamarie. 2012. *Orgasmology*. Next Wave: New Directions in Women's Studies. Duke University Press.

James, Sandy E., Jody L. Herman, Susan Rankin, Mara Keisling, Lisa Mottet, and Ma'ayan Anafi. 2016. "The Report of the 2015 US Transgender Survey." National Center for Transgender Equality.

James, Stanlie M. 1993. "Mothering: A Possible Black Feminist Link to Social Transformation?" In *Theorizing Black Feminisms: The Visionary Pragmatism of Black Women*, edited by Stanlie M. James and Abena P. A. Busia, 45. Routledge.

Jansen, Robert P. S. 2000. "Origin and Persistence of the Mitochondrial Genome." *Human Reproduction* 15 (Supplement 2): 1–10.

Janssen, Erick, and John Bancroft. 2007. "The Dual-Control Model: The Role of Sexual Inhibition and Excitation in Sexual Arousal and Behavior." In *The Psychophysiology of Sex*, edited by Eric Janssen, 197–222. Indiana University Press.

Jarne, Philippe, and Josh R. Auld. 2006. "Animals Mix It Up Too: The Distribution of Self-Fertilization Among Hermaphroditic Animals." *Evolution* 60 (9): 1816–24.

Jaron, Kamil S., Jens Bast, Reuben W. Nowell, T. Rhyker Ranallo-Benavidez, Marc Robinson-Rechavi, and Tanja Schwander. 2021. "Genomic Features of Parthenogenetic Animals." *Journal of Heredity* 112 (1): 19–33.

Jeffreys, Sheila. 1997. *The Spinster and Her Enemies: Feminism and Sexuality, 1880–1930*. Spinifex Press.

Johnson, Austin H. 2019. "Rejecting, Reframing, and Reintroducing: Trans People's Strategic Engagement with the Medicalisation of Gender Dysphoria." *Sociology of Health & Illness* 41 (3): 517–32.

Johnson, Eric Michael. 2021. "Ronald Fisher Is Not Being 'Cancelled,' But His Eugenic Advocacy Should Have Consequences." *This View of Life* (blog), April 12. https://thisviewoflife.com/ronald-fisher-is-not-being-cancelled-but-his-eugenic-advocacy-should-have-consequences/.

Johnston, Jill, and Lara Cushing. 2020. "Chemical Exposures, Health and Environmental Justice in Communities Living on the Fenceline of Industry." *Current Environmental Health Reports* 7 (1): 48–57.

Jones, Edward G., and Lorne M. Mendell. 1999. "Assessing the Decade of the Brain." *Science* 284 (5415): 739.

Jordan-Young, Rebecca M. 2010. *Brain Storm: The Flaws in the Science of Sex Differences*. Harvard University Press.

Jordan-Young, Rebecca M., and Raffaella I. Rumiati. 2012. "Hardwired for Sexism? Approaches to Sex/Gender in Neuroscience." *Neuroethics* 5: 305–15.

Kafer, Alison. 2013. *Feminist, Queer, Crip*. 1st ed. Indiana University Press.

Kafer, Alison. 2020. "Queer Disability Studies." In *The Cambridge Companion to Queer Studies*, edited by Siobhan B. Somerville, 93–107. Cambridge University Press.

Kagerer, Sabine, Tim Klucken, Sina Wehrum, et al. 2011. "Neural Activation Toward Erotic Stimuli in Homosexual and Heterosexual Males." *Journal of Sexual Medicine* 8 (11): 3132–43.

Kahan, Benjamin. 2013. *Celibacies: American Modernism and Sexual Life*. Duke University Press.

Kahan, Benjamin, and Greta LaFleur. 2023. "How to Do the History of Sexual Science." *GLQ: A Journal of Lesbian and Gay Studies* 29 (1): 1–12.

Kaiser Trujillo, Anelis, E. Ngubia Kessé, Oliver Rollins, Sergio Della Sala, and Roberto Cubelli. 2022. "A Discussion on the Notion of Race in Cognitive Neuroscience Research." *Cortex* 150 (May): 153–64.

Kaplan, Helen Singer. 1974. *The New Sex Therapy: Active Treatment of Sexual Dysfunctions*. Psychology Press.

Kaplan, Helen Singer. 1977. "Hypoactive Sexual Desire." *Journal of Sex & Marital Therapy* 3 (1): 3–9.

Kaplan, Helen Singer. 1979. *Disorders of Sexual Desire and Other New Concepts and Techniques in Sex Therapy*. Simon & Schuster.

Karama, Sherif, André Roch Lecours, Jean-Maxime Leroux, et al. 2002. "Areas of Brain Activation in Males and Females During Viewing of Erotic Film Excerpts." *Human Brain Mapping* 16 (1): 1–13.

Katz, Larry S. 2008. "Variation in Male Sexual Behavior." *Animal Reproduction Science* 105 (1–2): 64–71.

Kelleher, Sinéad, and Mike Murphy. 2022. "The Identity Development and Internalization of Asexual Orientation in Women: An Interpretative Phenomenological Analysis." *Sexual and Relationship Therapy* 39 (2): 359–89.

Kenney, Theresa N. 2020. "Thinking Asexually: Sapin-Sapin, Asexual Assemblages, and the Queer Possibilities of Platonic Relationalities." *Feminist Formations* 32 (3): 1–23.

Khan, Naveed A., and Ruqaiyyah Siddiqui. 2015. "Is There Evidence of Sexual Reproduction (Meiosis) in Acanthamoeba?" *Pathogens and Global Health* 109 (4): 193–95.

Kier, Bailey. 2010. "Interdependent Ecological Transsex: Notes on Re/Production, 'Transgender' Fish, and the Management of Populations, Species, and Resources." *Women & Performance: A Journal of Feminist Theory* 20 (3): 299–319.

Kilmann, Peter R., Joseph P. Boland, Shelley P. Norton, Edward Davidson, and Charlene Caid. 1986. "Perspectives of Sex Therapy Outcome: A Survey of AASECT Providers." *Journal of Sex & Marital Therapy* 12 (2): 116–38.

Kim, Eunjung. 2011. "Asexuality in Disability Narratives." *Sexualities* 14 (4): 479–93.

Kim, Eunjung. 2014. "Asexualities and Disabilities in Constructing Sexual Normalcy." In *Asexualities: Feminist and Queer Perspectives*, edited by K. J. Cerankowski and Megan Milks, 249–82. Routledge.

King, Hannah. 2021. "Asexuality 101: Isn't That for Plants?" *OpenLearn* (blog), March 9. https://www.open.edu/openlearn/society-politics-law/society-matters/society-politics-law/society-matters/asexuality-101-isnt-plants.

Kingsberg, Sheryl A. 2005. "The Testosterone Patch for Women." *International Journal of Impotence Research* 17 (5): 465–66.

Kinsey, Alfred, Wardell Pomeroy, Clyde Martin, and Paul Gebhard. 1953. *Sexual Behavior in the Human Female*. W. B. Saunders Company.

Kirksey, Eben. 2019. "Queer Love, Gender Bending Bacteria, and Life After the Anthropocene." *Theory, Culture & Society* 36 (6): 197–219.

Klein, Sanja, Onno Kruse, Charlotte Markert, Isabell Tapia León, Jana Strahler, and Rudolf Stark. 2020. "Subjective Reward Value of Visual Sexual Stimuli Is Coded in Human Striatum and Orbitofrontal Cortex." *Behavioural Brain Research* 393 (September): 112792.

Klesse, Christian. 2015. "Race and Sexology." *The International Encyclopedia of Human Sexuality*, edited by Patricia Whelehan and Anne Bolin, 1059–1114. Wiley-Blackwell.

Kli, Maria. 2018. "Eros and Thanatos: A Nondualistic Interpretation: The Dynamic of Drives in Personal and Civilizational Development from Freud to Marcuse." *Psychoanalytic Review* 105 (1): 67–89.

Klucken, Tim, Jan Schweckendiek, Christian J. Merz, et al. 2009. "Neural Activations of the Acquisition of Conditioned Sexual Arousal: Effects of Contingency Awareness and Sex." *Journal of Sexual Medicine* 6 (11): 3071–85.

Klucken, Tim, Sina Wehrum-Osinsky, Jan Schweckendiek, Onno Kruse, and Rudolf Stark. 2016. "Altered Appetitive Conditioning and Neural Connectivity in Subjects with Compulsive Sexual Behavior." *Journal of Sexual Medicine* 13 (4): 627–36.

Koedt, Anne. 1973. "The Myth of the Vaginal Orgasm." In *Radical Feminism*, edited by Anne Koedt, Ellen Levine, and Anita Rapone, 198–207. Quadrangle Books.

Koerth, Maggie. 2017. "The Complicated Legacy of a Panda Who Was Really Good at Sex." *FiveThirtyEight* (blog), November 28. https://fivethirtyeight.com /features/the-complicated-legacy-of-a-panda-who-was-really-good-at-sex/.

Kohiyama, Masamichi, Sota Hiraga, Ivan Matic, and Miroslav Radman. 2003. "Bacterial Sex: Playing Voyeurs 50 Years Later." *Science* 301 (5634): 802–3.

Komisaruk, Barry R. 2012. "A Scientist's Dilemma: Follow My Hypothesis or My Findings?" *Behavioural Brain Research*, Quo Vadis Behavioral Neuroscience: A Festschrift for Philip Teitelbaum, 231 (2): 262–65.

Komisaruk, Barry R., Nan J. Wise, Eleni Frangos, Wen-Ching Liu, Kachina Allen, and Stuart Brody. 2011. "Women's Clitoris, Vagina, and Cervix Mapped on the Sensory Cortex: fMRI Evidence." *Journal of Sexual Medicine* 8 (10): 2822–30.

Kono, T. 2006. "Genomic Imprinting Is a Barrier to Parthenogenesis in Mammals." *Cytogenetic and Genome Research* 113 (1–4): 31–35.

Korsgaard, Christine M. 2013. "Personhood, Animals, and the Law." *Think* 12 (34): 25–32.

Kościańska, Agnieszka. 2020. "Sexology." In *Companion to Sexuality Studies*, edited by Nancy A. Naples, 21–39. Wiley.

Krafft-Ebing, Richard. 1905. *Text Book of Insanity: Based on Clinical Observations for Practitioners and Students of Medicine*, translated by Charles Gilbert Chaddock. F. A. Davis Company.

Krafft-Ebing, Richard. 1906. *Psychopathia Sexualis: With Especial Reference to the Antipathic Sexual Instinct, a Medico-Forensic Study*. Rebman.

Kragel, Philip A., Marianne C. Reddan, Kevin S. LaBar, and Tor D. Wager. 2019. "Emotion Schemas Are Embedded in the Human Visual System." *Science Advances* 5 (7): eaaw4358.

Kreukels, Baudewijntje P. C., and Antonio Guillamon. 2016. "Neuroimaging Studies in People with Gender Incongruence." *International Review of Psychiatry* 28 (1): 120–28.

Kriegeskorte, Nikolaus, Martin A. Lindquist, Thomas E. Nichols, Russell A. Poldrack, and Edward Vul. 2010. "Everything You Never Wanted to Know About Circular Analysis, but Were Afraid to Ask." *Journal of Cerebral Blood Flow and Metabolism* 30 (9): 1551–57.

Kringelbach, Morten L., and Kent C. Berridge. 2016. "Neuroscience of Reward, Motivation, and Drive." In *Recent Developments in Neuroscience Research on Human Motivation*, by Sung-il Kim, Johnmarshall Reeve, and Mimi Bong, 23–35. Emerald Group Publishing.

Kühn, Simone, and Jürgen Gallinat. 2011. "A Quantitative Meta-Analysis on Cue-Induced Male Sexual Arousal." *Journal of Sexual Medicine* 8 (8): 2269–75.

Kurowicka, Anna. 2014. "What Can Asexuality Do for Queer Theories?" *LES Online* 6 (1): 21–27.

Kurowicka, Anna. 2023. "Contested Intersections: Asexuality and Disability, Illness, or Trauma." *Sexualities* 28 (1–2). https://doi.org/10.1177/13634607231170781.

Kurowicka, Anna, and Ela Przybylo. 2020. "Polish Asexualities: Catholic Religiosity and Asexual Online Activisms in Poland." In *LGBTQ+ Activism in Central and Eastern Europe: Resistance, Representation and Identity*, edited by Radzhana Buyantueva and Maryna Shevtsova, 289–311. Springer International Publishing.

Labuski, Christine. 2015. *It Hurts Down There: The Bodily Imaginaries of Female Genital Pain*. SUNY Press.

ladypoetess. 2011. "Meant for Vegetarians, Works for Aces Too!" *Asexuality* (blog). Livejournal, March 16. https://asexuality.livejournal.com/819541.html.

Lake, James A. 2011. "Lynn Margulis (1938–2011)." *Nature* 480 (7378): 458.

Laplanche, Jean, and J. B. Pontalis. 1973. *The Language of Psycho-Analysis*. Hogarth Press.

Laqueur, Thomas. 2004. *Solitary Sex: A Cultural History of Masturbation*. Zone Books.

Larson, Edward J. 2010. "Biology and the Emergence of the Anglo-American Eugenics Movement." In *Biology and Ideology from Descartes to Dawkins*, edited by Denis R. Alexander and Ronald L. Numbers, 165–91. University of Chicago Press.

Leca, Jean-Baptiste, Noelle Gunst, Lydia Ottenheimer Carrier, and Paul L. Vasey. 2014. "Inter-Group Variation in Non-Conceptive Sexual Activity in Female Japanese Macaques: Could It Be Cultural?" *Animal Behavior and Cognition* 1 (3): 387–409.

Lee, Min Joo. 2020. "Intimacy Beyond Sex: Korean Television Dramas, Nonsexual Masculinities, and Transnational Erotic Desires." *Feminist Formations* 32 (3): 100–120.

Lee, Robyn, and Roxanne Mykitiuk. 2018. "Surviving Difference: Endocrine-Disrupting Chemicals, Intergenerational Justice and the Future of Human Reproduction." *Feminist Theory* 19 (2): 205–21.

Lee, S. W., B. S. Jeong, J. Choi, and J.-W. Kim. 2015. "Sex Differences in Interactions Between Nucleus Accumbens and Visual Cortex by Explicit Visual Erotic Stimuli: An fMRI Study." *International Journal of Impotence Research* 27 (5): 161–66.

Le Espiritu, Yen. 2004. "All Men Are Not Created Equal: Asian Men in US History." In *Men's Lives*, edited by Michael S. Kimmel and Michael A. Messner, 21–27. Oxford University Press.

Lehtonen, Jussi, Michael D. Jennions, and Hanna Kokko. 2012. "The Many Costs of Sex." *Trends in Ecology & Evolution* 27 (3): 172–78.

Lehtonen, Jussi, Hanna Kokko, and Geoff A. Parker. 2016. "What Do Isogamous Organisms Teach Us About Sex and the Two Sexes?" *Philosophical Transactions of the Royal Society B: Biological Sciences* 371 (1706): 20150532.

Leiblum, Sandra R., and Raymond C. Rosen, eds. 1988. *Sexual Desire Disorders*. Guilford Press.

Lenormand, Thomas, Jan Engelstädter, Susan E. Johnston, Erik Wijnker, and Christoph R. Haag. 2016. "Evolutionary Mysteries in Meiosis." *Philosophical Transactions: Biological Sciences* 371 (1706): 1–14.

Leonard, Janet L. 2018. "The Evolution of Sexual Systems in Animals." In *Transitions Between Sexual Systems: Understanding the Mechanisms of, and Pathways Between, Dioecy, Hermaphroditism and Other Sexual Systems*, edited by Janet L. Leonard, 1–58. Springer International Publishing.

Levine, Stephen B. 2003. "The Nature of Sexual Desire: A Clinician's Perspective." *Archives of Sexual Behavior* 32 (3): 279–85.

Lewis, Carolyn. 2010. *Prescription for Heterosexuality: Sexual Citizenship in the Cold War Era*. 1st ed. University of North Carolina Press.

Li, Wei. 2021. "Asexual Species Can Adapt to Their Environment Even Without Changing Their DNA." *Science in the News* (blog), September 27. https://sitn.hms.harvard.edu/flash/2021/asexual-species-can-adapt-to-their-environment-even-without-changing-their-dna/.

Liddell, Jessica L., and Sarah G. Kington. 2021. "'Something Was Attacking Them and Their Reproductive Organs': Environmental Reproductive Justice in an Indigenous Tribe in the United States Gulf Coast." *International Journal of Environmental Research and Public Health* 18 (2): 666.

Lief, Harold I. 1977. "Inhibited Sexual Desire." *Medical Aspects of Human Sexuality* 7 (1): 94–95.

Lin, Kai. 2017. "The Medicalization and Demedicalization of Kink: Shifting Contexts of Sexual Politics." *Sexualities* 20 (3): 302–23.

Lindzey, Jonathan, and David Crews. 1992. "Individual Variation in Intensity of Sexual Behaviors in Captive Male *Cnemidophorus inornatus*." *Hormones and Behavior* 26 (1): 46–55.

Livingston, Tyler N., Peter O. Rerick, and Deborah Davis. 2024. "Race/Ethnicity and Relationship Stereotypes in Child Sex Abuse Cases." *Psychological Reports* 127 (1): 112–23.

Llaveria Caselles, Eric. 2021. "Epistemic Injustice in Brain Studies of (Trans)Gender Identity." *Frontiers in Sociology* 6 (March): 608328.

Lloyd, Elisabeth A. 2005. *The Case of the Female Orgasm: Bias in the Science of Evolution*. Harvard University Press.

Long, Xipeng, Fangfang Tian, Yushan Zhou, Bochao Cheng, and Zhiyun Jia. 2020. "Different Neural Correlates of Sexually Preferred and Sexually Nonpreferred Stimuli." *Journal of Sexual Medicine* 17 (7): 1254–67.

Long, Xipeng, Fangfang Tian, Yushan Zhou, Bochao Cheng, Siqi Yi, and Zhiyun Jia. 2019. "The Neural Correlates of Sexual Arousal and Sexual Disgust." SSRN Scholarly Paper. https://dx.doi.org/10.2139/ssrn.3458493.

Lopez, Cristina S. 2001. "Post-Biomechanics: Difference and Gender in Margulis and Sagan's What Is Sex?" *POROI* 1 (1): 7, 98–121.

Lorber, Judith. 2021. *The New Gender Paradox: Fragmentation and Persistence of the Binary*. John Wiley & Sons.

Lorde, Audre. 1984. *Sister Outsider: Essays and Speeches*. 1st ed. Crossing Press.

Lorenzi, M. C., and G. Sella. 2008. "A Measure of Sexual Selection in Hermaphroditic Animals: Parentage Skew and the Opportunity for Selection." *Journal of Evolutionary Biology* 21 (3): 827–33.

Ludwig, David, Inkeri Koskinen, Zinhle Mncube, Luana Poliseli, and Luis Reyes-Galindo. 2021. *Global Epistemologies and Philosophies of Science*. Routledge.

Lugones, María. 2007. "Heterosexualism and the Colonial/Modern Gender System." *Hypatia* 22 (1): 186–219.

Lugones, María. 2010. "Toward a Decolonial Feminism." *Hypatia* 25 (4): 742–59.

Lund, Emily M., and Bayley A. Johnson. 2014. "Asexuality and Disability: Strange but Compatible Bedfellows." *Sexuality and Disability* 33 (1): 123–32.

Lykins, Amy D., Marta Meana, and Gregory P. Strauss. 2008. "Sex Differences in Visual Attention to Erotic and Non-Erotic Stimuli." *Archives of Sexual Behavior* 37 (2): 219–28.

MacKenzie, Steven. 2020. "Parthenogenesis." *The Gale Encyclopedia of Science* 6 (November): 3322–24.

MacNeela, Pádraig, and Aisling Murphy. 2015. "Freedom, Invisibility, and Community: A Qualitative Study of Self-Identification with Asexuality." *Archives of Sexual Behavior* 44 (3): 799–812.

Malatino, Hil. 2020. *Trans Care*. University of Minnesota Press.

Malihi, Zarintaj A., Janet L. Fanslow, Ladan Hashemi, Pauline J. Gulliver, and Tracey K. D. McIntosh. 2021. "Prevalence of Nonpartner Physical and Sexual Violence Against People with Disabilities." *American Journal of Preventive Medicine* 61 (3): 329–37.

Malón, Agustin. 2012. "Pedophilia: A Diagnosis in Search of a Disorder." *Archives of Sexual Behavior* 41 (5): 1083–97.

Mamo, Laura. 2007. *Queering Reproduction: Achieving Pregnancy in the Age of Technoscience*. 1st ed. Duke University Press.

Mamo, Laura, and Eli Alston-Stepnitz. 2015. "Queer Intimacies and Structural Inequalities: New Directions in Stratified Reproduction." *Journal of Family Issues* 36 (4): 519–40.

Mancini, Elena. 2010. *Magnus Hirschfeld and the Quest for Sexual Freedom: A History of the First International Sexual Freedom Movement*. Springer.

Mandegar, Mohammad A., and Sarah P. Otto. 2007. "Mitotic Recombination Counteracts the Benefits of Genetic Segregation." *Proceedings of the Royal Society B: Biological Sciences* 274 (1615): 1301–7.

Manevski, Darko. 2021. "Pandas Living in the Lap of Luxury 'Too Lazy' for Sex." *Newsweek*, October 3.

Margolin, Leslie. 2023a. "The Third Backdoor: How the DSM Casebooks Pathologized Homosexuality." *Journal of Homosexuality* 70 (2): 291–306.

Margolin, Leslie. 2023b. "Why Is Absent/Low Sexual Desire a Mental Disorder (Except When Patients Identify as Asexual)?" *Psychology & Sexuality* 14 (4): 720–33.

Margulis, Lynn, and Dorion Sagan. 1990. *Origins of Sex: Three Billion Years of Genetic Recombination.* Yale University Press.

Marshall, Barbara L. 2006. "The New Virility: Viagra, Male Aging and Sexual Function." *Sexualities* 9 (3): 345–62.

Marshall, Barbara L., and Stephen Katz. 2013. "From Androgyny to Androgens: Resexing the Aging Body." In *Age Matters: Realigning Feminist Thinking*, edited by Toni M. Calasanti and Kathleen F. Slevin, 75–97. Routledge.

Martens, Koen, Hugh D. Loxdale, and Isa Schön. 2009. "The Elusive Clone—In Search of Its True Nature and Identity." In *Lost Sex: The Evolutionary Biology of Parthenogenesis*, edited by Isa Schön, Koen Martens, and Peter van Dijk, 187–200. Springer.

Martin-Wintle, Meghan S., David Shepherdson, Guiquan Zhang, et al. 2015. "Free Mate Choice Enhances Conservation Breeding in the Endangered Giant Panda." *Nature Communications* 6 (1): 10125.

Martynova, Olga V., Galina V. Portnova, and I. Yu. Orlov. 2017. "An fMRI Study of the Emotional Perception of Erotic Images in Men Aged 49–74 Years." *Neuroscience and Behavioral Physiology* 47 (4): 393–401.

Masters, William H., and Virginia E. Johnson. 1970. *Human Sexual Inadequacy.* Little Brown.

Matthew-Onabanjo, A., M. Bjurlin, and M. Rogers. 2024. "(319) Evaluating the Racial Distribution of Clinical Trial Participants for Drugs Leading to FDA Approval of Novel Biologics for the Treatment of Hypoactive Sexual Desire Disorder (HSDD)." *Journal of Sexual Medicine* 21 (Supplement 1): qdae001.304.

McCall, Leslie. 2005. "The Complexity of Intersectionality." *Signs* 30 (3): 1771–1800.

McCormack, Karen. 2005. "Stratified Reproduction and Poor Women's Resistance." *Gender & Society* 19 (5): 660–79.

McKittrick, Katherine. 2015. *Sylvia Wynter: On Being Human as Praxis.* Duke University Press.

McLaren, Angus. 1999. *Twentieth-Century Sexuality: A History.* Wiley-Blackwell.

McLaren, Angus. 2007. *Impotence: A Cultural History.* University of Chicago Press.

McWhorter, Ladelle. 1997. "Foucault's Attack on Sex-Desire." *Philosophy Today* 41 (1): 160–65.

Meana, Marta. 2010. "Elucidating Women's (Hetero)Sexual Desire: Definitional Challenges and Content Expansion." *Journal of Sex Research* 47 (2–3): 104–22.

Meirmans, Stephanie. 2009. "The Evolution of the Problem of Sex." In *Lost Sex: The Evolutionary Biology of Parthenogenesis*, edited by Isla Schön, Koen Martens, and Peter van Dijk, 21–46. Springer.

Meirmans, Stephanie, Patrick G. Meirmans, and Lawrence R. Kirkendall. 2012. "The Costs of Sex: Facing Real-World Complexities." *Quarterly Review of Biology* 87 (1): 19–40.

Meirmans, Stephanie, and Roger Strand. 2010. "Why Are There So Many Theories for Sex, and What Do We Do with Them?" *Journal of Heredity* 101 (Supplement 1): S3–12.

Metzger, Coraline Danielle, Ulf Eckert, Johann Steiner, et al. 2010. "High Field fMRI Reveals Thalamocortical Integration of Segregated Cognitive and Emotional Processing in Mediodorsal and Intralaminar Thalamic Nuclei." *Frontiers in Neuroanatomy* 4: 138.

Meyer, Axel. 2010. "George C. Williams (1926–2010)." *Nature* 467 (7317): 790.

Meyerowitz, Joanne. 2002. *How Sex Changed: A History of Transsexuality in the United States.* 1st ed. Harvard University Press.

Michael, R. P., and G. S. Saayman. 1967. "Individual Differences in the Sexual Behaviour of Male Rhesus Monkeys (*Macaca mulatta*) Under Laboratory Conditions." *Animal Behaviour* 15 (4): 460–66.

Michod, Richard. 1996. *Eros and Evolution: A Natural Philosophy of Sex.* Basic Books.

Milani, Sonia, Jia Yu Zhang, Bozena Zdaniuk, Anthony Bogaert, Gerulf Rieger, and Lori A. Brotto. 2023. "Examining Visual Attention Patterns Among Asexual and Heterosexual Individuals." *Journal of Sex Research* 60 (2): 271–81.

Miller, Nicholas E. 2017. "Asexuality and Its Discontents: Making the 'Invisible Orientation' Visible in Comics." *Inks: The Journal of the Comics Studies Society* 1 (3): 354–76.

Mintzes, Barbara, Leonore Tiefer, and Lisa Cosgrove. 2021. "Bremelanotide and Flibanserin for Low Sexual Desire in Women: The Fallacy of Regulatory Precedent." *Drug and Therapeutics Bulletin* 59 (12): 185–88.

Mitchell, Heather, and Gwen Hunnicutt. 2019. "Challenging Accepted Scripts of Sexual 'Normality': Asexual Narratives of Non-Normative Identity and Experience." *Sexuality & Culture* 23 (2): 507–24.

Mitchell, Kirstin R., Kaye A. Wellings, and Cynthia Graham. 2014. "How Do Men and Women Define Sexual Desire and Sexual Arousal?" *Journal of Sex & Marital Therapy* 40 (1): 17–32.

Mitricheva, Ekaterina, Rui Kimura, Nikos K. Logothetis, and Hamid R. Noori. 2019. "Neural Substrates of Sexual Arousal Are Not Sex Dependent." *Proceedings of the National Academy of Sciences* 116 (31): 15671–76.

Moccia, Lorenzo, Marianna Mazza, Marco Di Nicola, and Luigi Janiri. 2018. "The Experience of Pleasure: A Perspective Between Neuroscience and Psychoanalysis." *Frontiers in Human Neuroscience* 12 (September 4): 359.

Moldowan, Patrick D., Ronald J. Brooks, and Jacqueline D. Litzgus. 2020. "Sex, Shells, and Weaponry: Coercive Reproductive Tactics in the Painted Turtle, *Chrysemys picta.*" *Behavioral Ecology and Sociobiology* 74 (12): 142.

Mollet, Amanda L. 2020. "'I Have a Lot of Feelings, Just None in the Genitalia Region': A Grounded Theory of Asexual College Students' Identity Journeys." *Journal of College Student Development* 61 (2): 189–206.

Mollet, Amanda L., and Wayne Black. 2023. "Coercive Rape Tactics Perpetrated Against Asexual College Students: A Quantitative Analysis Considering Students' Multiple Identities." *Journal of College Student Development* 64 (1): 96–101.

Mollet, Amanda L., and Brian Lackman. 2019. "Asexual Student Invisibility and Erasure in Higher Education." In *Rethinking LGBTQIA Students and Collegiate Contexts: Identity, Policies, and Campus Climate*, edited by Eboni M. Zamani-Gallaher, Devika Dibya Choudhuri, and Jason L. Taylor, 78–98. Routledge.

Mollet, Amanda L., and Brian Lackman. 2021. "Allonormativity and Compulsory Sexuality." In *Encyclopedia of Queer Studies in Education*, edited by Kamden K. Strunk and Stephanie Anne Shelton, 26–30. Brill.

Monk, Julia D., Erin Giglio, Ambika Kamath, Max R. Lambert, and Caitlin E. McDonough. 2019. "An Alternative Hypothesis for the Evolution of Same-Sex Sexual Behaviour in Animals." *Nature Ecology & Evolution* 3 (12): 1622–31.

Moore, Alison. 2009. "Frigidity, Gender and Power in French Cultural History." *French Cultural Studies* 20 (4): 331–49.

Moore, Alison, and Peter Cryle. 2010. "Frigidity at the Fin de Siècle in France: A Slippery and Capacious Concept." *Journal of the History of Sexuality* 19 (2): 243–61.

Moradigaravand, Danesh, and Jan Engelstädter. 2013. "The Evolution of Natural Competence: Disentangling Costs and Benefits of Sex in Bacteria." *American Naturalist* 182 (4): E112–E126.

Moran, Nancy A., and Tyler Jarvik. 2010. "Lateral Transfer of Genes from Fungi Underlies Carotenoid Production in Aphids." *Science* 328 (5978): 624–27.

Morgan, Joan. 2015. "Why We Get Off: Moving Towards a Black Feminist Politics of Pleasure." *Black Scholar* 45 (4): 36–46.

Morgensen, Scott Lauria. 2010. "Settler Homonationalism: Theorizing Settler Colonialism Within Queer Modernities." *GLQ: A Journal of Lesbian and Gay Studies* 16 (1–2): 105–31.

Morgensen, Scott Lauria. 2012. "Theorising Gender, Sexuality and Settler Colonialism: An Introduction." *Settler Colonial Studies* 2 (2): 2–22.

Morgensen, Scott Lauria. 2021. "Heteronormativity." In *Keywords for Gender and Sexuality Studies*, edited by the Keywords Feminist Editorial Collective. New York University Press.

Morris, Margaret Kissam. 2002. "Audre Lorde: Textual Authority and the Embodied Self." *Frontiers: A Journal of Women Studies* 23 (1): 168–88.

Morton, Timothy. 2010. "Guest Column: Queer Ecology." *PMLA* 125 (2): 273–82.

Moser, Charles. 2010. "Hypersexual Disorder: Just More Muddled Thinking." *Archives of Sexual Behavior* 40 (2): 227–29.

Mott, Maryann. 2006. "Panda 'Rent' Too High, U.S. Zoos Say." *National Geographic*, March 13. https://www.nationalgeographic.com/animals/article/united-states-panda-zoo-animals.

Moulier, Virginie, Harold Mouras, Mélanie Pélégrini-Issac, et al. 2006. "Neuroanatomical Correlates of Penile Erection Evoked by Photographic Stimuli in Human Males." *NeuroImage* 33 (2): 689–99.

Mouras, Harold, Serge Stoléru, Jacques Bittoun, et al. 2003. "Brain Processing of Visual Sexual Stimuli in Healthy Men: A Functional Magnetic Resonance Imaging Study." *NeuroImage* 20 (2): 855–69.

Murphy, Michelle. 2017. "Alterlife and Decolonial Chemical Relations." *Cultural Anthropology* 32 (4): 494–503.

Murphy, Sara Flannery. 2021. *Girl One*. MCD.

Musser, Amber Jamilla. 2012. "On the Orgasm of the Species: Female Sexuality, Science and Sexual Difference." *Feminist Review* 102 (1): 1–20.

Nakamura, Gabriel, Bruno Eleres Soares, Valério D. Pillar, José Alexandre Felizola Diniz-Filho, and Leandro Duarte. 2023. "Three Pathways to Better Recognize the Expertise of Global South Researchers." *npj Biodiversity* 2 (1): 1–4.

Narra, Hema Prasad, and Howard Ochman. 2006. "Of What Use Is Sex to Bacteria?" *Current Biology* 16 (17): R705–10.

Nash, Jennifer C. 2014. *The Black Body in Ecstasy: Reading Race, Reading Pornography*. Duke University Press.

Neal, Jennifer K., and Juli Wade. 2007. "Androgen Receptor Expression and Morphology of Forebrain and Neuromuscular Systems in Male Green Anoles Displaying Individual Differences in Sexual Behavior." *Hormones and Behavior* 52 (2): 228–36.

Neuhaus, Jessamyn. 2000. "The Importance of Being Orgasmic: Sexuality, Gender, and Marital Sex Manuals in the United States, 1920–1963." *Journal of the History of Sexuality* 9 (4): 447–73.

Nguyen, Tan Hoang. 2014. *A View from the Bottom: Asian American Masculinity and Sexual Representation*. Duke University Press.

Nicholls, Henry. 2010. *The Way of the Panda: The Curious History of China's Political Animal*. Profile.

Nichols, Ben. 2020. "Reproductive." In *Same Old: Queer Theory, Literature and the Politics of Sameness*, edited by Ben Nichols, 1–38. Manchester University Press.

Nichols, Margaret. 2004. "Lesbian Sexuality/Female Sexuality: Rethinking 'Lesbian Bed Death.'" *Sexual and Relationship Therapy* 19 (4): 363–71.

Nieuwenhuis, Bart P. S., and Timothy Y. James. 2016. "The Frequency of Sex in Fungi." *Philosophical Transactions: Biological Sciences* 371 (1706): 1–12.

Obermann, Kyle. 2021. "China Declares Pandas No Longer Endangered—but Threats Persist." *National Geographic*, September 1. https://www.nationalgeographic.com/animals/article/pandas-are-off-chinas-endangered-list-but-threats-persist.

O'Connor, Cliodhna, Geraint Rees, and Helene Joffe. 2012. "Neuroscience in the Public Sphere." *Neuron* 74 (2): 220–26.

Ogawa, Kota, and Toru Miura. 2014. "Aphid Polyphenisms: Trans-Generational Developmental Regulation Through Viviparity." *Frontiers in Physiology* 5 (January): 1–11.

Okami, Paul, and Todd K. Shackelford. 2001. "Human Sex Differences in Sexual Psychology and Behavior." *Annual Review of Sex Research* 12 (1): 186–241.

Okasha, Samir. 2016. *Philosophy of Science: A Very Short Introduction*. Oxford University Press.

O'Laughlin, Logan Natalie. 2020. "Troubling Figures: Endocrine Disruptors, Intersex Frogs, and the Logics of Environmental Science." *Catalyst: Feminism, Theory, Technoscience* 6 (1): 1–29.

Oosterhuis, Harry. 2000. *Stepchildren of Nature: Krafft-Ebing, Psychiatry, and the Making of Sexual Identity*. University of Chicago Press.

Orive, Maria E. 2020. "The Evolution of Sex." In *The Theory of Evolution: Principles, Concepts, and Assumptions*, edited by Samuel M. Schneiner and David P. Mindell, 273–95. University of Chicago Press.

Ortigue, Stephanie, and Francesco Bianchi-Demicheli. 2008. "The Chronoarchitecture of Human Sexual Desire: A High-Density Electrical Mapping Study." *NeuroImage* 43 (2): 337–45.

Otto, Sarah P. 2008. "Sexual Reproduction and the Evolution of Sex." *Nature Education* 1 (1): 182.

Otto, Sarah P. 2009. "The Evolutionary Enigma of Sex." *American Naturalist* 174 (S1): S1–14.

Otto, Sarah P., and Thomas Lenormand. 2002. "Resolving the Paradox of Sex and Recombination." *Nature Reviews Genetics* 3 (4): 252–62.

Owen, Ianna Hawkins. 2014. "On the Racialization of Asexuality." In *Asexualities: Feminist and Queer Perspsectives*, edited by K. J. Cerankowski and Megan Milks, 119–35. Routledge.

Owen, Ianna Hawkins. 2018. "Still, Nothing: Mammy and Black Asexual Possibility." *Feminist Review* 120 (1): 70–84.

Owen, Ianna Hawkins. 2022. "More: Cake, Feedism, and Asexuality." *Social Text* 40 (2 [151]): 93–111.

Palmer, Seth. 2014. "Asexual Inverts and Sexual Perverts: Locating the Sarimbavy of Madagascar Within Fin-de-Siècle Sexological Theories." *TSQ: Transgender Studies Quarterly* 1 (3): 368–86.

Pamuk, Zeynep. 2021. *Politics and Expertise: How to Use Science in a Democratic Society*. Princeton University Press.

Pappas, Stephanie. 2015. "The Key to Making Baby Pandas? Love." Livescience.com, December 18. https://www.livescience.com/53158-key-to-panda-mating.html.

Parada, Mayte, Marina Gérard, Kevin Larcher, Alain Dagher, and Yitzchak M. Binik. 2016. "Neural Representation of Subjective Sexual Arousal in Men and Women." *Journal of Sexual Medicine* 13 (10): 1508–22.

Parisi, Luciana. 2004. *Abstract Sex: Philosophy, Bio-Technology and the Mutations of Desire*. 1st ed. Continuum.

Parker, Emily Anne. 2018. "The Human as Double Bind: Sylvia Wynter and the Genre of 'Man.'" *Journal of Speculative Philosophy* 32 (3): 439–49.

Parmenter, Joshua G., Renee V. Galliher, and Adam D. A. Maughan. 2021. "LGBTQ+ Emerging Adults Perceptions of Discrimination and Exclusion Within the LGBTQ+ Community." *Psychology & Sexuality* 12 (4): 289–304.

Patil, Vrushali. 2022. *Webbed Connectivities: The Imperial Sociology of Sex, Gender, and Sexuality*. University of Minnesota Press.

Paul, Thomas, Boris Schiffer, Thomas Zwarg, et al. 2008. "Brain Response to Visual Sexual Stimuli in Heterosexual and Homosexual Males." *Human Brain Mapping* 29 (6): 726–35.

Paus, Tomáš. 2010. "Population Neuroscience: Why and How." *Human Brain Mapping* 31 (6): 891–903.

Pavia, Will. 2013. "Forget Su Doku, Orgasms Are Better for the Brain." *The Times* (London), August 5. https://www.thetimes.co.uk/article/forget-su-doku-orgasms-are-better-for-the-brain-16csxdwnwk7.

Pawlowski, Jan, and Fabien Burki. 2009. "Untangling the Phylogeny of Amoeboid Protists." *Journal of Eukaryotic Microbiology* 56 (1): 16–25.

Peiretti-Courtis, Delphine. 2021. "African Hypersexuality: A Threat to White Settlers? The Stigmatization of 'Black Sexuality' as a Means of Regulating 'White Sexuality.'" In *Histories of Sexology: Between Science and Politics*, edited by Alain Giami and Sharman Levinson, 263–76. Springer International Publishing.

Perret, Meg. 2020. "'Chemical Castration': White Genocide and Male Extinction in Rhetoric of Endocrine Disruption." *NiCHE* (blog), World, June 9. https://niche-canada.org/2020/06/09/chemical-castration-white-genocide-and-male-extinction-in-rhetoric-of-endocrine-disruption/.

Persson, Sofia, and Madeleine Pownall. 2021. "Can Open Science Be a Tool to Dismantle Claims of Hardwired Brain Sex Differences? Opportunities and Challenges for Feminist Researchers." *Psychology of Women Quarterly* 45 (4): 493–504.

Peterson, Zoë D., and Erick Janssen. 2007. "Ambivalent Affect and Sexual Response: The Impact of Co-Occurring Positive and Negative Emotions on Subjective and Physiological Sexual Responses to Erotic Stimuli." *Archives of Sexual Behavior* 36 (6): 793–807.

Pettit, Michael. 2012. "The Queer Life of a Lab Rat." *History of Psychology* 15 (3): 217.

Pfaus, James G. 2009. "Reviews: Pathways of Sexual Desire." *Journal of Sexual Medicine* 6 (6): 1506–33.

Pfaus, James G., Tod E. Kippin, and Genaro Coria-Avila. 2003. "What Can Animal Models Tell Us About Human Sexual Response?" *Annual Review of Sex Research* 14 (1): 1–63.

Pfaus, James G., Tina Scardochio, Mayte Parada, Christine Gerson, Gonzalo R. Quintana, and Genaro A. Coria-Avila. 2016. "Do Rats Have Orgasms?" *Socioaffective Neuroscience & Psychology* 6 (October): 10.3402/snp.v6.31883.

Philip, Marion. 2019. "Frigidity, Curses, and Imagination: Thinking the Absence of Male Desire in the Seventeenth and Eighteenth Centuries." *French History and Civilization* 8 (January): 15–33.

Phoenix, Charles H., and Kathleen C. Chambers. 1988. "Testosterone Therapy in Young and Old Rhesus Males That Display Low Levels of Sexual Activity." *Physiology & Behavior* 43 (4): 479–84.

Piel, Helen. 2017. "Local Heroes: John Maynard Smith: (1920–2004): A Good 'Puzzle-Solver' with an 'Accidental Career.'" *Science Blog from the British Library* (blog), March 15. https://blogs.bl.uk/science/2017/03/local-heroes-john-maynard-smith-1920-2004-a-good-puzzle-solver-with-an-accidental-career.html.

Poeppl, Timm B., Berthold Langguth, Rainer Rupprecht, et al. 2016. "The Neural Basis of Sex Differences in Sexual Behavior: A Quantitative Meta-Analysis." *Frontiers in Neuroendocrinology* 43 (October): 28–43.

Poeppl, Timm B., Katrin Sakreida, and Simon B. Eickhoff. 2020. "Neural Substrates of Sexual Arousal Revisited: Dependent on Sex." *Proceedings of the National Academy of Sciences* 117 (21): 11204–5.

Pollock, Anne. 2016. "Queering Endocrine Disruption." In *Object-Oriented Feminism*, edited by Katherine Behar, 183–99. University of Minnesota Press.

Ponseti, Jorge, Hartmut A. Bosinski, Stephan Wolff, et al. 2006. "A Functional Endophenotype for Sexual Orientation in Humans." *NeuroImage* 33 (3): 825–33.

Ponseti, Jorge, Oliver Granert, Olav Jansen, et al. 2009. "Assessment of Sexual Orientation Using the Hemodynamic Brain Response to Visual Sexual Stimuli." *Journal of Sexual Medicine* 6 (6): 1628–34.

Ponseti, Jorge, Oliver Granert, Olav Jansen, et al. 2012. "Assessment of Pedophilia Using Hemodynamic Brain Response to Sexual Stimuli." *Archives of General Psychiatry* 69 (2): 187–94.

Portillo, Wendy, Edwards Antonio-Cabrera, Francisco J. Camacho, Néstor Fabián Díaz, and Raúl G. Paredes. 2013. "Behavioral Characterization of Non-Copulating Male Mice." *Hormones and Behavior* 64 (1): 70–80.

Portillo, Wendy, Elizabeth Basañez, and Raúl G. Paredes. 2003. "Permanent Changes in Sexual Behavior Induced by Medial Preoptic Area Kindling-Like Stimulation." *Brain Research* 961 (1): 10–14.

Portillo, Wendy, Claudia G. Castillo, Socorro Retana-Márquez, Charles E. Roselli, and Raúl G. Paredes. 2007. "Neuronal Activity of Aromatase Enzyme in Non-Copulating Male Rats." *Journal of Neuroendocrinology* 19 (2): 139–41.

Portillo, Wendy, Néstor Fabián Díaz, Edwards Antonio Cabrera, Alonso Fernández-Guasti, and Raúl G. Paredes. 2006. "Comparative Analysis of Immunoreactive Cells for Androgen Receptors and Oestrogen Receptor α in Copulating and Non-Copulating Male Rats." *Journal of Neuroendocrinology* 18 (3): 168–76.

Portillo, Wendy, and Raúl G. Paredes. 2003. "Sexual and Olfactory Preference in Noncopulating Male Rats." *Physiology & Behavior* 80 (1): 155–62.

Portillo, Wendy, and Raúl G. Paredes. 2019. "Motivational Drive in Non-Copulating and Socially Monogamous Mammals." *Frontiers in Behavioral Neuroscience* 13 (October 4). https://doi.org/10.3389/fnbeh.2019.00238.

Poskett, James. 2022a. *Horizons: A Global History of Science.* Penguin Books.

Poskett, James. 2022b. *Materials of the Mind: Phrenology, Race, and the Global History of Science, 1815–1920.* University of Chicago Press.

Potts, Annie. 2000. "'The Essence of the Hard On': Hegemonic Masculinity and the Cultural Construction of 'Erectile Dysfunction.'" *Men and Masculinities* 3 (1): 85–103.

Potts, Annie. 2022. "For Women's Pleasure? Interspecies Sexual Violence and Gynocentric Sex Research." In *Feminist Animal Studies*, edited by Erika Cudworth, Ruth E. McKie, and Di Turgoose, 134–148. Routledge.

Prause, Nicole, and Carla Harenski. 2014. "Inhibition, Lack of Excitation, or Suppression: fMRI Pilot of Asexuality." In *Asexualities: Feminist and Queer Perspectives*, edited by K. J. Cerankowski and Megan Milks, 35–54. Routledge.

Pruitt-Young, Sharon. 2021. "Finally Some Good News! China Says Giant Pandas Are No Longer Endangered." NPR, July 9, sec. Animals. https://www.npr.org /2021/07/09/1014593425/china-giant-pandas-endangered-vulnerable-iucn.

Przybylo, Ela. 2012. "Producing Facts: Empirical Asexuality and the Scientific Study of Sex." *Feminism & Psychology*, April.

Przybylo, Ela. 2014. "Masculine Doubt and Sexual Wonder." In *Asexualities: Feminist and Queer Perspectives*, edited by K. J. Cerankowski and Megan Milks, 225–46. Routledge.

Przybylo, Ela. 2019. *Asexual Erotics: Intimate Readings of Compulsory Sexuality*. Ohio State University Press.

Przybylo, Ela. 2022. "Asexuality and Compulsory Sexuality." In *The Palgrave Encyclopedia of Sexuality Education*, edited by Louisa Allen and Mary Lou Rasmussen. 1–10. Springer International Publishing.

Przybylo, Ela, and Danielle Cooper. 2014. "Asexual Resonances Tracing a Queerly Asexual Archive." *GLQ: A Journal of Lesbian and Gay Studies* 20 (3): 297–318.

Puar, Jasbir K. 2017. *The Right to Maim: Debility, Capacity, Disability*. ANIMA: Critical Race Studies Otherwise. Duke University Press.

Public Face of Science Initiative. 2018. "Perceptions of Science in America: A Report from the Public Face of Science Initiative." American Academy of Arts and Sciences.

Putkinen, Vesa, Sanaz Nazari-Farsani, Tomi Karjalainen, et al. 2022. "Accurate Sex Classification from Neural Responses to Sexual Stimuli." *bioRxiv* (January 11). https://doi.org/10.1101/2022.01.10.473972.

Quammen, David. 2018. *The Tangled Tree: A Radical New History of Life*. 1st ed. Simon & Schuster.

"Queerplatonic Relationship." n.d. LGBTQIA+ Wiki. Accessed September 25, 2023. https://lgbtqia.fandom.com/wiki/Queerplatonic_relationship.

Raghavan, Val. 2005. *Double Fertilization: Embryo and Endosperm Development in Flowering Plants*. Springer.

Randell, John B. 1959. "Transvestitism and Trans-Sexualism. A Study of 50 Cases." *British Medical Journal* 2 (5164): 1448–52.

"R/Asexuality—What Animal Would You Choose to Represent Asexuality?" n.d. Reddit. Accessed December 13, 2022. https://www.reddit.com/r/asexuality /comments/pyun17/what_animal_would_you_choose_to_represent/.

Rauch, Scott L., Lisa M. Shin, Darin D. Dougherty, et al. 1999. "Neural Activation during Sexual and Competitive Arousal in Healthy Men." *Psychiatry Research: Neuroimaging* 91 (1): 1–10.

Redfield, Rosemary J. 2001. "Do Bacteria Have Sex?" *Nature Reviews Genetics* 2 (8): 634–40.

Redouté, Jérôme, Serge Stoléru, Marie-Claude Grégoire, et al. 2000. "Brain Processing of Visual Sexual Stimuli in Human Males." *Human Brain Mapping* 11 (3): 162–77.

Reidpath, Daniel D., and Pascale Allotey. 2019. "The Problem of 'Trickle-Down Science' from the Global North to the Global South." *BMJ Global Health* 4 (4): e001719.

Reis, Elizabeth. 2009. *Bodies in Doubt: An American History of Intersex*. JHU Press.

Reisner, Sari L., Soon Kyu Choi, Jody L. Herman, Walter Bockting, Evan A. Krueger, and Ilan H. Meyer. 2023. "Sexual Orientation in Transgender Adults in the United States." *BMC Public Health* 23 (1): 1799.

Renaud, Patrice, Christian Joyal, Serge Stoléru, Mathieu Goyette, Nikolaus Weiskopf, and Niels Birbaumer. 2011. "Real-Time Functional Magnetic Imaging-Brain-Computer Interface and Virtual Reality: Promising Tools for the Treatment of Pedophilia." *Progress in Brain Research* 192:263–72.

Rensenbrink, Greta. 2010. "Parthenogenesis and Lesbian Separatism: Regenerating Women's Community Through Virgin Birth in the United States in the 1970s and 1980s." *Journal of the History of Sexuality* 19 (2): 288–316.

Rich, Adrienne. 1980. "Compulsory Heterosexuality and Lesbian Existence." *Signs* 5 (4): 631–60.

Richardson, Aaron O., and Jeffrey D. Palmer. 2007. "Horizontal Gene Transfer in Plants." *Journal of Experimental Botany* 58 (1): 1–9.

Richardson, Diane. 1997. "Sexuality and Feminism." In *Introducing Women's Studies: Feminist Theory and Practice*, edited by Victoria Robinson and Diane Richardson, 152–74. Macmillan Education UK.

Riddle, Dorothy I. 2014. "Evolving Notions of Nonhuman Personhood: Is Moral Standing Sufficient?" *Journal of Ethics and Emerging Technologies* 24 (3): 4–19.

Rifkin, Mark. 2012. *The Erotics of Sovereignty: Queer Native Writing in the Era of Self-Determination*. University of Minnesota Press.

Rippon, Gina, Rebecca Jordan-Young, Anelis Kaiser, and Cordelia Fine. 2014. "Recommendations for Sex/Gender Neuroimaging Research: Key Principles and Implications for Research Design, Analysis, and Interpretation." *Frontiers in Human Neuroscience* 8 (August 27). https://doi.org/10.3389/fnhum.2014.00650.

RiverOfLife, Martuwarra, Alessandro Pelizzon, Anne Poelina, et al. 2021. "Yoongoorrookoo." *Griffith Law Review* 30 (3): 505–29.

Rkasnuam, Dawy. 2019. "Asian and Asexual: How I Came to Own My Asexuality While Fighting Cultural Stereotypes." *The Body Is Not an Apology*, May 2. https://thebodyisnotanapology.com/magazine/asexualtiy-and-cultural-stereotypes/ (site discontinued).

Roach, Mary. 2009. *Bonk: The Curious Coupling of Science and Sex*. Reprint ed. W. W. Norton & Company.

Robbins, Nicolette K., Kathryn Graff Low, and Anna N. Query. 2015. "A Qualitative Exploration of the 'Coming Out' Process for Asexual Individuals." *Archives of Sexual Behavior* 45 (3): 751–60.

Robinson, Paul A. 1976. *The Modernization of Sex: Havelock Ellis, Alfred Kinsey, William Masters, and Virginia Johnson*. 1st ed. Harper & Row.

Robitzski, Dan. 2022. "How a Grasshopper Gave Up Sex, Took Up Cloning." *The Scientist Magazine*, October 31. https://www.the-scientist.com/how-a-grasshopper-gave-up-sex-took-up-cloning-70582.

Rochon, Christopher, Gonzalo Otazu, Isaac L. Kurtzer, Randy F. Stout, and Raddy L. Ramos. 2019. "Quantitative Indicators of Continued Growth in Undergraduate Neuroscience Education in the US." *Journal of Undergraduate Neuroscience Education* 18 (1): A51–56.

Rodríguez-Rocha, Vivian. 2021. "Social Reproduction Theory: State of the Field and New Directions in Geography." *Geography Compass* 15 (8): e12586.

Rolin, Kristina. 2021. "Philosophy of Science Analytic Feminist Approaches." In *The Oxford Handbook of Feminist Philosophy*, edited by Kim Q. Hall and Ásta, 226–36. Oxford University Press.

Roll-Hansen, Nils. 2010. "Eugenics and the Science of Genetics." In *The Oxford Handbook of the History of Eugenics*, edited by Philippa Levine and Alison Bashford, 80–97. Oxford University Press.

Roselli, Charles E. 2018. "Neurobiology of Gender Identity and Sexual Orientation." *Journal of Neuroendocrinology* 30 (7): 1–8.

Roselli, Charles E., and Fred Stormshak. 2009. "The Neurobiology of Sexual Partner Preferences in Rams." *Hormones and Behavior* 55 (5): 611–20.

Roselli, Charles E., Fred Stormshak, John N. Stellflug, and John A. Resko. 2002. "Relationship of Serum Testosterone Concentrations to Mate Preferences in Rams." *Biology of Reproduction* 67 (1): 263–68.

Rosenberg, Gabriel. 2017. "How Meat Changed Sex: The Law of Interspecies Intimacy After Industrial Reproduction." *GLQ: A Journal of Lesbian and Gay Studies* 23 (4): 473–507.

Rosenberg, Jordana. 2014. "The Molecularization of Sexuality: On Some Primitivisms of the Present." *Theory & Event* 17 (2): https://muse.jhu.edu/article/546470.

Roth, Lateefah, Peer Briken, and Johannes Fuss. 2023. "Masturbation in the Animal Kingdom." *Journal of Sex Research* 60 (6): 786–98.

Rubin, David A. 2017. *Intersex Matters: Biomedical Embodiment, Gender Regulation, and Transnational Activism*. SUNY Press.

Rubin, Gayle. 1984. "Thinking Sex: Notes for a Radical Theory of the Politics of Sexuality." In *Pleasure and Danger: Exploring Female Sexuality*, edited by Carole S. Vance. Routledge and Kegan Paul.

Rudy, Kathy. 2001. "Radical Feminism, Lesbian Separatism, and Queer Theory." *Feminist Studies* 27 (1): 191–222.

Ruesink, Gerben B., and Janniko R. Georgiadis. 2017. "Brain Imaging of Human Sexual Response: Recent Developments and Future Directions." *Current Sexual Health Reports* 9 (4): 183–91.

Russ, Joanna. 1975. *The Female Man*. Bantam.

Ruti, Mari. 2015. *The Age of Scientific Sexism: How Evolutionary Psychology Promotes Gender Profiling and Fans the Battle of the Sexes*. Bloomsbury Publishing USA.

Sabatinelli, Dean, Margaret M. Bradley, Peter J. Lang, Vincent D. Costa, and Francesco Versace. 2007. "Pleasure Rather than Salience Activates Human Nucleus Accumbens and Medial Prefrontal Cortex." *Journal of Neurophysiology* 98 (3): 1374–79.

Safron, Adam, Bennett Barch, J. Michael Bailey, Darren R. Gitelman, Todd B. Parrish, and Paul J. Reber. 2007. "Neural Correlates of Sexual Arousal in Homosexual and Heterosexual Men." *Behavioral Neuroscience* 121 (2): 237–48.

Safron, Adam, David Sylva, Victoria Klimaj, et al. 2017. "Neural Correlates of Sexual Orientation in Heterosexual, Bisexual, and Homosexual Men." *Scientific Reports* 7 (1): 1–15.

Samarrai, Fariss. 2009. "Sex and the Single Snail: Study Shows Benefits of Sexual Reproduction over Asexual." *UVA Today* (blog), December 17. https://news .virginia.edu/content/sex-and-single-snail-study-shows-benefits-sexual -reproduction-over-asexual.

Sample, Ian. 2011. "Female Orgasm Captured in Series of Brain Scans." *The Guardian*, November 14, sec. Science. https://www.theguardian.com/science/2011 /nov/14/female-orgasm-recorded-brain-scans.

Sandford, Stella. 2004. "Let's Talk About Sex: Book Review of: *Abstract Sex: Philosophy, Bio-Technology and the Mutations of Desire* by Luciana Parisi, and *The Sex Appeal of the Inorganic*, by Mario Perniola." *Radical Philosophy* 127:35–40.

Sandford, Stella. 2019. "A Thousand Tiny 'Sexes' or None?" *Sluice Magazine* (Autumn-Winter): 36–41.

Saunders, Rebecca. 2020. *Bodies of Work: The Labour of Sex in the Digital Age*. Springer Nature.

Sawicki, Jana. 1991. *Disciplining Foucault: Feminism, Power, and the Body*. Routledge.

Sawicki, Jana. 2010. "Foucault, Queer Theory, and the Discourse of Desire." In *Foucault and Philosophy*, edited by Timothy O'Leary and Christopher Falzon, 185–203. John Wiley & Sons.

Schaefer, Donovan. 2018. "Precopulatory Sexual Cannibalism and Other Accidents: Evolution, Material Trans Theory, and Natural Law." *Studies in Gender & Sexuality* 19 (1): 28–35.

Schaefer, Donovan. 2021. "Darwin's Orchids: Evolution, Natural Law, and the Diversity of Desire." *GLQ: A Journal of Lesbian and Gay Studies* 27 (4): 525–50.

Schalk, Sami. 2018. *Bodyminds Reimagined: (Dis)ability, Race, and Gender in Black Women's Speculative Fiction*. Duke University Press.

Scherrer, Kristin S. 2008. "Coming to an Asexual Identity: Negotiating Identity, Negotiating Desire." *Sexualities* 11 (5): 621–41.

Schiebinger, Londa L. 1993. *Nature's Body: Gender in the Making of Modern Science*. Beacon Press.

Schmidt, Thomas M. 2002. "Bacteria and Archaea." In *Encyclopedia of Evolution*, edited by Mark Pagel. Oxford University Press. Accessed May 29, 2025. https:// www-oxfordreference-com.wake.idm.oclc.org/view/10.1093/acref /9780195122008.001.0001/acref-9780195122008-e-36.

Schmitz, Sigrid, and Grit Hoppner. 2014. *Gendered Neurocultures: Feminist and Queer Perspectives on Current Brain Discourses*. Zaglossus.

Schön, Isa, Dunja K. Lamatsch, and Koen Martens. 2008. "Lessons to Learn from Ancient Asexuals." In *Recombination and Meiosis: Models, Means, and Evolution*, edited by Richard Egel and Dirk-Henner Lankenau, 341–76. Springer.

Schön, Isa, and Koen Martens. 2018. "Paradox of Sex." In *Oxford Bibliographies in Evolutionary Biology*, edited by Noah Whiteman. Oxford University Press.

Schön, Isa, Koen Martens, and Peter van Dijk, eds. 2009. *Lost Sex: The Evolutionary Biology of Parthenogenesis*. Springer.

Schuller, Kyla. 2018. *The Biopolitics of Feeling: Race, Sex, and Science in the Nineteenth Century*. Duke University Press.

Schwander, Tanja. 2016. "Evolution: The End of an Ancient Asexual Scandal." *Current Biology* 26 (6): R233–35.

Schwander, Tanja, and Bernard J. Crespi. 2009. "Twigs on the Tree of Life? Neutral and Selective Models for Integrating Macroevolutionary Patterns with Microevolutionary Processes in the Analysis of Asexuality." *Molecular Ecology* 18 (1): 28–42.

Schwander, Tanja, Lee Henry, and Bernard J. Crespi. 2011. "Molecular Evidence for Ancient Asexuality in Timema Stick Insects." *Current Biology* 21 (13): 1129–34.

Schwander, Tanja, and Benjamin P. Oldroyd. 2016. "Androgenesis: Where Males Hijack Eggs to Clone Themselves." *Philosophical Transactions: Biological Sciences* 371 (1706): 1–13.

Scott, Joan W. 1988. "Deconstructing Equality-Versus-Difference: Or, the Uses of Poststructuralist Theory for Feminism." *Feminist Studies* 14 (1): 33–50.

Scott, Susie, and Matt Dawson. 2015. "Rethinking Asexuality: A Symbolic Interactionist Account." *Sexualities* 18 (1–2): 3–19.

Sedgwick, Eve Kosofsky. 1990. *Epistemology of the Closet*. University of California Press.

Sedgwick, Eve Kosofsky. 1993. *Tendencies*. Duke University Press.

Seitler, Dana. 2003. "Unnatural Selection: Mothers, Eugenic Feminism, and Charlotte Perkins Gilman's Regeneration Narratives." *American Quarterly* 55 (1): 61–88.

Seo, Younghee, Bumseok Jeong, Ji-Woong Kim, and Jeewook Choi. 2009. "Plasma Concentration of Prolactin, Testosterone Might Be Associated with Brain Response to Visual Erotic Stimuli in Healthy Heterosexual Males." *Psychiatry Investigation* 6 (3): 194–203.

Seok, Ji-Woo, Mi-Sook Park, and Jin-Hun Sohn. 2016. "Neural Pathways in Processing of Sexual Arousal: A Dynamic Causal Modeling Study." *International Journal of Impotence Research* 28 (5): 184–88.

Seok, Ji-Woo, and Jin-Hun Sohn. 2015. "Neural Substrates of Sexual Desire in Individuals with Problematic Hypersexual Behavior." *Frontiers in Behavioral Neuroscience* 9 (November 29). https://doi.org/10.3389/fnbeh.2015.00321.

Serano, Julia. 2008. "A Matter of Perspective: A Transsexual Woman-Centric Critique of Dreger's 'Scholarly History' of the Bailey Controversy." *Archives of Sexual Behavior* 37 (3): 491–94.

Serano, Julia. 2010. "The Case Against Autogynephilia." *International Journal of Transgenderism* 12 (3): 176–87.

Serano, Julia. 2020. "Autogynephilia: A Scientific Review, Feminist Analysis, and Alternative 'Embodiment Fantasies' Model." *Sociological Review* 68 (4): 763–78.

Sescousse, Guillaume, Guillaume Barbalat, Philippe Domenech, and Jean-Claude Dreher. 2013. "Imbalance in the Sensitivity to Different Types of Rewards in Pathological Gambling." *Brain* 136 (8): 2527–38.

Sescousse, Guillaume, Xavier Caldú, Bàrbara Segura, and Jean-Claude Dreher. 2013. "Processing of Primary and Secondary Rewards: A Quantitative Meta-Analysis and Review of Human Functional Neuroimaging Studies." *Neuroscience and Biobehavioral Reviews* 37 (4): 681–96.

Sescousse, Guillaume, Yansong Li, and Jean-Claude Dreher. 2015. "A Common Currency for the Computation of Motivational Values in the Human Striatum." *Social Cognitive and Affective Neuroscience* 10 (4): 467–73.

Sescousse, Guillaume, Jérôme Redouté, and Jean-Claude Dreher. 2010. "The Architecture of Reward Value Coding in the Human Orbitofrontal Cortex." *Journal of Neuroscience* 30 (39): 13095–104.

Seymour, Nicole. 2018. *Bad Environmentalism: Irony and Irreverence in the Ecological Age*. University of Minnesota Press.

Shakespeare, Tom. 1996. "Power and Prejudice: Issue of Gender, Sexuality and Disability." In *Disability and Society*, edited by Len Barton, 191–214. Routledge.

Shakespeare, Tom. 2000. "Disabled Sexuality: Toward Rights and Recognition." *Sexuality and Disability* 18 (3): 159–66.

Shek, Yen Ling. 2007. "Asian American Masculinity: A Review of the Literature." *Journal of Men's Studies* 14 (3): 379–91.

Shildrick, Margrit. 2007. "Dangerous Discourses: Anxiety, Desire, and Disability." *Studies in Gender and Sexuality* 8 (3): 221–44.

Shuster, Stef M. 2021. *Trans Medicine: The Emergence and Practice of Treating Gender*. NYU Press.

Siggy. 2022. "A Whirlwind History of Asexual Communities." *The Asexual Agenda* (blog), August 7. https://asexualagenda.wordpress.com/2022/08/07/a-whirlwind-history-of-asexual-communities/.

Sillitoe, Paul. 2009. *Local Science vs. Global Science: Approaches to Indigenous Knowledge in International Development*. Berghahn Books.

Sinclair, Rebekah. 2020. "Exploding Individuals: Engaging Indigenous Logic and Decolonizing Science." *Hypatia* 35 (1): 58–74.

Singh, Nilanchali, Pallavi Sharma, and Neha Mishra. 2020. "Female Sexual Dysfunction: Indian Perspective and Role of Indian Gynecologists." *Indian Journal of Community Medicine* 45 (3): 333–37.

SisterSong. n.d. "Reproductive Justice." Accessed June 9, 2023. https://www.sistersong.net/reproductive-justice.

Skorska, Malvina N., Morag A. Yule, Anthony F. Bogaert, and Lori A. Brotto. 2023. "Patterns of Genital and Subjective Sexual Arousal in Cisgender Asexual Men." *Journal of Sex Research* 60 (2): 253–70.

Slonczewski, Joan. 1986. *A Door into Ocean*. Arbor House.

Smith, Justin. 2020. "'[T]he Happiest, Well-Feddest Wolf in Harlem': Asexuality as Resistance to Social Reproduction in Claude McKay's *Home to Harlem*." *Feminist Formations* 32 (3): 51–74.

Smithers, Gregory D. 2017. *Science, Sexuality, and Race in the United States and Australia, 1780–1940*. University of Nebraska Press.

Snaza, Nathan. 2020. "Asexuality and Erotic Biopolitics." *Feminist Formations* 32 (3): 121–44.

Snoeren, Eelke M. S., and Anders Ågmo. 2014. "The Incentive Value of Males' 50-kHz Ultrasonic Vocalizations for Female Rats (Rattus norvegicus)." *Journal of Comparative Psychology* 128 (1): 40–55.

Snoeren, Eelke M. S., Johnny S. W. Chan, Trynke R. de Jong, Marcel D. Waldinger, Berend Olivier, and Ronald S. Oosting. 2011. "A New Female Rat Animal Model for Hypoactive Sexual Desire Disorder: Behavioral and Pharmacological Evidence." *Journal of Sexual Medicine* 8 (1): 44–56.

Somerville, Siobhan B. 2000. *Queering the Color Line: Race and the Invention of Homosexuality in American Culture*. Duke University Press.

Song, Yixian, B. Drossel, and S. Scheu. 2011. "Tangled Bank Dismissed Too Early." *Oikos* 120 (11): 1601–7.

Songster, E. Elena. 2018. *Panda Nation: The Construction and Conservation of China's Modern Icon*. Oxford University Press.

Spector, Ilana P., Michael P. Carey, and Lynne Steinberg. 1996. "The Sexual Desire Inventory: Development, Factor Structure, and Evidence of Reliability." *Journal of Sex & Marital Therapy* 22 (3): 175–90.

Speijer, Dave, Julius Lukeš, and Marek Eliáš. 2015. "Sex Is a Ubiquitous, Ancient, and Inherent Attribute of Eukaryotic Life." *Proceedings of the National Academy of Sciences* 112 (29): 8827–34.

Spillers, Hortense J. 1987. "Mama's Baby, Papa's Maybe: An American Grammar Book." *Diacritics* 17 (2): 65–83.

Sprinkle, Robert Hunt, and Devon C. Payne-Sturges. 2021. "Mixture Toxicity, Cumulative Risk, and Environmental Justice in United States Federal Policy, 1980–2016." *Environmental Health* 20 (1): 104.

Spurgas, Alyson K. 2012. "Where Is My Subjectivity? Techno-Imagery, Femininity & Desire—Social Text." *SocialText Online*, April. https://socialtextjournal.org/periscope_article/where-is-my-subjectivity-techno-imagery-femininity-desire/.

Spurgas, Alyson K. 2020. *Diagnosing Desire: Biopolitics and Femininity into the Twenty-First Century*. 1st ed. Ohio State University Press.

Spurgas, Alyson K. 2022. "The Feminization of 'Responsive' Desire." In *Introducing the New Sexuality Studies*, edited by Nancy L. Fischer, Laurel Westbrook, and Steven Seidman, 401–9. 4th ed. Routledge.

Stark, Rudolf, Sanja Klein, Onno Kruse, et al. 2019. "No Sex Difference Found: Cues of Sexual Stimuli Activate the Reward System in Both Sexes." *Neuroscience* 416 (September): 63–73.

Stark, Rudolf, Anne Schienle, Cornelia Girod, et al. 2005. "Erotic and Disgust-Inducing Pictures—Differences in the Hemodynamic Responses of the Brain." *Biological Psychology* 70 (1): 19–29.

Stein, Edward. 1999. *The Mismeasure of Desire: The Science, Theory and Ethics of Sexual Orientation*. Ideologies of Desire. Oxford University Press.

Stekel, Wilhelm. 1926. *Frigidity in Woman in Relation to Her Love Life*. Translated by James S. Van Teslaar. Boni and Liveright.

Stoléru, Serge. 2014. "Reading the Freudian Theory of Sexual Drives from a Functional Neuroimaging Perspective." *Frontiers in Human Neuroscience* 8 (March 17). https://doi.org/10.3389/fnhum.2014.00157.

Stoléru, Serge, Véronique Fonteille, Christel Cornélis, Christian Joyal, and Virginie Moulier. 2012. "Functional Neuroimaging Studies of Sexual Arousal and Orgasm in Healthy Men and Women: A Review and Meta-Analysis." *Neuroscience & Biobehavioral Reviews* 36 (6): 1481–1509.

Stoléru, Serge, Marie-Claude Grégoire, Daniel Gérard, et al. 1999. "Neuroanatomical Correlates of Visually Evoked Sexual Arousal in Human Males." *Archives of Sexual Behavior* 28 (1): 1–21.

Stoléru, Serge, Jérôme Redouté, Nicolas Costes, et al. 2003. "Brain Processing of Visual Sexual Stimuli in Men with Hypoactive Sexual Desire Disorder." *Psychiatry Research: Neuroimaging* 124 (2): 67–86.

Stone, Aaron J. 2023. "Toward a Black Vernacular Sexology." GLQ: *A Journal of Lesbian and Gay Studies* 29 (1): 27–42.

Strahler, Jana, Onno Kruse, Sina Wehrum-Osinsky, Tim Klucken, and Rudolf Stark. 2018. "Neural Correlates of Gender Differences in Distractibility by Sexual Stimuli." *NeuroImage* 176 (August): 499–509.

Sturgeon, Noël. 2010. "Penguin Family Values: The Nature of Planetary Environmental Reproductive Justice." In *Queer Ecologies: Sex, Nature, Politics, Desire*, edited by Catriona Mortimer-Sandilands and Bruce Erickson, 102–33. Indiana University Press.

Subramaniam, Banu. 2014. *Ghost Stories for Darwin: The Science of Variation and the Politics of Diversity*. University of Illinois Press.

Subramaniam, Banu. 2016. "Stories We Tell: Feminism, Science, Methodology." *Economic and Political Weekly* 51 (18): 57–63.

Subramaniam, Banu, Laura Foster, Sandra Harding, Deboleena Roy, and Kim TallBear. 2016. "Feminism, Postcolonialism, Technoscience." In *The Handbook of Science and Technology Studies*, edited by Rayvon Fouché, Clark A. Miller, Ulrike Felt, and Laurel Smith-Doerr, 407–34. 4th ed. MIT Press.

Suess, Amets, Karine Espineira, and Pau Crego Walters. 2014. "Depathologization." TSQ: *Transgender Studies Quarterly* 1 (1–2): 73–77.

Sukel, Kayt. 2011. "I Had an Orgasm in an MRI Scanner." *The Guardian*, November 16, sec. Science. https://www.theguardian.com/science/blog/2011/nov/16/orgasm-mri-scanner.

Sukel, Kayt. 2012. "Orgasm Research: Climax in an MRI Machine? Been There, Done That." *HuffPost*, January 9, sec. Women. https://www.huffpost.com/entry/orgasm_b_1193191.

Sukel, Kayt. 2013. *This Is Your Brain on Sex: The Science Behind the Search for Love.* Illustrated ed. Simon & Schuster.

Sulloway, Frank. 1992. *Freud, Biologist of the Mind: Beyond the Psychoanalytic Legend.* Harvard University Press.

Sunobe, Tomoki, Tetsuya Sado, Kiyoshi Hagiwara, et al. 2017. "Evolution of Bidirectional Sex Change and Gonochorism in Fishes of the Gobiid Genera *Trimma*, *Priolepis*, and *Trimmatom*." *Science of Nature* 104 (3): 15.

Sussman, Robert Wald. 2016. *The Myth of Race: The Troubling Persistence of an Unscientific Idea.* Reprint ed. Harvard University Press.

Sylva, David, Adam Safron, A. M. Rosenthal, Paul J. Reber, Todd B. Parrish, and J. Michael Bailey. 2013. "Neural Correlates of Sexual Arousal in Heterosexual and Homosexual Women and Men." *Hormones and Behavior* 64 (4): 673–84.

Takebayashi, Naoki, and Peter L. Morrell. 2001. "Is Self-Fertilization an Evolutionary Dead End? Revisiting an Old Hypothesis with Genetic Theories and a Macroevolutionary Approach." *American Journal of Botany* 88 (7): 1143–50.

Talia. 2019. "The Biology Definition of Asexuality May Have More Impact on the Sexual Orientation Than We Think." *Asexual Agenda* (blog), June 25. https://asexualagenda.wordpress.com/2019/06/25/the-biology-definition-of-asexuality-may-have-more-impact-on-the-sexual-orientation-than-we-think/.

TallBear, Kim. 2022. "Making Love and Relations Beyond Settler Sex and Family." In *Queerly Canadian: An Introductory Reader in Sexuality Studies*, edited by Scott Rayter and Laine Halpern Zisman, 18–28. 2nd ed. Women's Press.

Tata, Elvis Fon, Kenneth Anchang Yongabi, Ernest Dzelamonyuy, Noela Ijang Forbang, and Edi Achuh Geh. 2023. "Decolonizing African Science: Efforts, Challenges, and Future Directions." *International Journal of Science and Research Archive* 10 (1): 823–33.

Taylor, Chloë. 2024. "Introduction: Why Gender and Animals?" In *The Routledge Companion to Gender and Animals*, edited by Chloë Taylor. Routledge.

Tepper, Mitchell S. 2000. "Sexuality and Disability: The Missing Discourse of Pleasure." *Sexuality and Disability* 18 (4): 283–90.

Terry, Jennifer. 1999. *An American Obsession: Science, Medicine, and Homosexuality in Modern Society.* 1st ed. University of Chicago Press.

Terry, Jennifer. 2000. "'Unnatural Acts' in Nature: The Scientific Fascination with Queer Animals." *GLQ: A Journal of Lesbian and Gay Studies* 6 (2): 151–93.

"Testosterone Patches for Female Sexual Dysfunction." 2009. *Drug and Therapeutics Bulletin* 47 (3): 30–34.

Thornhill, Randy, and Craig Palmer. 2000. *A Natural History of Rape: Biological Bases of Sexual Coercion.* MIT Press.

Tiefer, Leonore. 2004. *Sex Is Not a Natural Act & Other Essays.* 2nd ed. Westview Press.

Tiefer, Leonore. 2006. "Female Sexual Dysfunction: A Case Study of Disease Mongering and Activist Resistance." *PLoS Med* 3 (4): e178.

Toates, Frederick. 2009. "An Integrative Theoretical Framework for Understanding Sexual Motivation, Arousal, and Behavior." *Journal of Sex Research* 46 (2–3): 168–93.

tommy92. 2014. "(Indirect) Mentions of Asexuality in Magnus Hirschfeld's Books." Asexual Visibility and Education Network, February 8. https://www.asexuality.org/en/topic/98639-indirect-mentions-of-asexuality-in-magnus-hirschfelds-books/.

Tóth, András, Lóránt Székvölgyi, and Tibor Vellai. 2022. "The Genome Loading Model for the Origin and Maintenance of Sex in Eukaryotes." *Biologia Futura* 73 (4): 345–57.

Tripp, Erin A. 2016. "Is Asexual Reproduction an Evolutionary Dead End in Lichens?" *Lichenologist* 48 (5): 559–80.

Trivers, R. L. 1972. "Parental Investment and Sexual Selection." In *Sexual Selection and the Descent of Man*, edited by Bernard G. Campbell, 136–79. 1st ed. Routledge.

Troisi, Alfonso, and Monica Carosi. 1998. "Female Orgasm Rate Increases with Male Dominance in Japanese Macaques." *Animal Behaviour* 56 (5): 1261–66.

Tsukahara, Shinji, Moeko Kanaya, and Korehito Yamanouchi. 2014. "Neuroanatomy and Sex Differences of the Lordosis-Inhibiting System in the Lateral Septum." *Frontiers in Neuroscience* 8 (September): 299.

Tuana, Nancy. 2004. "Coming to Understand: Orgasm and the Epistemology of Ignorance." *Hypatia* 19 (1): 194–232.

Tuana, Nancy. 2021. "Feminist New Materialisms." In *The Oxford Handbook of Feminist Philosophy*, edited by Kim Q. Hall and Ásta, 385–94. Oxford University Press.

Turner, Crystasany. 2024. "Othermothering: A Black Feminist History of Communal Early Childhood Education in America." *Journal of African American Women and Girls in Education* 4 (2): 108–27.

Uddin, Lisa. 2013. "Panda Gardens and Public Sex at the National Zoological Park." *Public*, no. 41 (November): 81–92.

Vance, Carter. 2018. "Towards a Historical Materialist Concept of Asexuality and Compulsory Sexuality." *Studies in Social Justice* 12 (1): 133–51.

Vandermassen, Griet. 2005. *Who's Afraid of Charles Darwin? Debating Feminism and Evolutionary Theory.* Rowman & Littlefield Publishers.

Vandermassen, Griet. 2011. "Evolution and Rape: A Feminist Darwinian Perspective." *Sex Roles* 64 (9–10): 732–47.

van Dijk, Peter. 2009. "Apomixis: Basics for Non-Botanists." In *Lost Sex: The Evolutionary Biology of Parthenogenesis*, edited by Isa Schön, Koen Martens, and Peter van Dijk, 47–62. Springer.

van Lankveld, Jacques J. D. M., and Fren T. Y. Smulders. 2008. "The Effect of Visual Sexual Content on the Event-Related Potential." *Biological Psychology* 79 (2): 200–208.

van't Hof, Sophie R., and Nicoletta Cera. 2021. "Specific Factors and Methodological Decisions Influencing Brain Responses to Sexual Stimuli in Women." *Neuroscience & Biobehavioral Reviews* 131 (December): 164–78.

van't Hof, Sophie R., Lukas Van Oudenhove, Erick Janssen, et al. 2022. "The Brain Activation-Based Sexual Image Classifier (BASIC): A Sensitive and Specific fMRI Activity Pattern for Sexual Image Processing." *Cerebral Cortex* 32 (14): 3014–30.

Vares, Tiina. 2018. "'My [Asexuality] Is Playing Hell with My Dating Life': Romantic Identified Asexuals Negotiate the Dating Game." *Sexualities* 21 (4): 520–36.

Vasey, Paul L. 2002. "Same-Sex Sexual Partner Preference in Hormonally and Neurologically Unmanipulated Animals." *Annual Review of Sex Research* 13 (December): 141–79.

Ventura-Aquino, Elisa, and Raúl G. Paredes. 2017. "Animal Models in Sexual Medicine: The Need and Importance of Studying Sexual Motivation." *Sexual Medicine Reviews* 5 (1): 5–19.

Verhoeven, Koen J. F., and Veronica Preite. 2014. "Epigenetic Variation in Asexually Reproducing Organisms." *Evolution* 68 (3): 644–55.

Wallace-Sanders, Kimberly. 2008. *Mammy: A Century of Race, Gender, and Southern Memory*. University of Michigan Press.

Wallis, Robert J. 2009. "Re-Enchanting Rock Art Landscapes: Animic Ontologies, Nonhuman Agency and Rhizomic Personhood." *Time and Mind* 2 (1): 47–69.

Walter, Martin, Felix Bermpohl, Harold Mouras, et al. 2008. "Distinguishing Specific Sexual and General Emotional Effects in fMRI—Subcortical and Cortical Arousal During Erotic Picture Viewing." *NeuroImage* 40 (4): 1482–94.

Walters, Suzanna Danuta. 2014. *The Tolerance Trap: How God, Genes, and Good Intentions Are Sabotaging Gay Equality*. NYU Press.

Walz, Thomas. 2002. "Crones, Dirty Old Men, Sexy Seniors: Representations of the Sexuality of Older Persons." *Journal of Aging and Identity* 7 (2): 99–112.

Wang, Thelma. 2022. "Trans as Brain Intersex: The Trans-Intersex Nexus in Neurobiological Research." *TSQ: Transgender Studies Quarterly* 9 (2): 172–83.

Ward, Anna E. 2018. "Between the Screens: Brain Imaging, Pornography, and Sex Research." *Catalyst: Feminism, Theory, Technoscience* 4 (1): 1–28.

Ward, Caleb. 2023. "Audre Lorde's Erotic as Epistemic and Political Practice." *Hypatia* 38 (4): 896–917.

Ward, Jane. 2015. *Not Gay: Sex Between Straight White Men*. NYU Press.

Warner, Michael, ed. 1993. *Fear of a Queer Planet: Queer Politics and Social Theory*. University of Minnesota Press.

Warren, Wesley C., Raquel García-Pérez, Sen Xu, et al. 2018. "Clonal Polymorphism and High Heterozygosity in the Celibate Genome of the Amazon Molly." *Nature Ecology & Evolution* 2 (4): 669–79.

Watkinson, Sarah C., Lynne Boddy, and Nicholas Money. 2015. *The Fungi*. Academic Press.

Weaver, Harlan. 2021. *Bad Dog: Pit Bull Politics and Multispecies Justice*. University of Washington Press.

Weaver, Sara, and Carla Fehr. 2017. "Values, Practices, and Metaphysical Assumptions in the Biological Sciences." In *The Routledge Companion to Feminist Philosophy*, edited by Ann Garry, Serene J. Khader, and Alison Stone, 314–28. Routledge.

Wehrum, Sina, Tim Klucken, Sabine Kagerer, et al. 2013. "Gender Commonalities and Differences in the Neural Processing of Visual Sexual Stimuli." *Journal of Sexual Medicine* 10 (5): 1328–42.

Wehrum-Osinsky, Sina, Tim Klucken, Sabine Kagerer, Bertram Walter, Andrea Hermann, and Rudolf Stark. 2014. "At the Second Glance: Stability of Neural Responses Toward Visual Sexual Stimuli." *Journal of Sexual Medicine* 11 (11): 2720–37.

Wei, Fuwen, Yibo Hu, Li Yan, Yonggang Nie, Qi Wu, and Zejun Zhang. 2015. "Giant Pandas Are Not an Evolutionary Cul-de-Sac: Evidence from Multidisciplinary Research." *Molecular Biology and Evolution* 32 (1): 4–12.

Wei, Fuwen, Ronald Swaisgood, Yibo Hu, et al. 2015. "Progress in the Ecology and Conservation of Giant Pandas." *Conservation Biology* 29 (6): 1497–1507.

Wei, Yanchang, Cai-Rong Yang, and Zhen-Ao Zhao. 2022. "Viable Offspring Derived from Single Unfertilized Mammalian Oocytes." *Proceedings of the National Academy of Sciences of the United States of America* 119 (12): e2115248119.

Weinbaum, Alys Eve. 2001. "Writing Feminist Genealogy: Charlotte Perkins Gilman, Racial Nationalism, and the Reproduction of Maternalist Feminism." *Feminist Studies* 27 (2): 271–302.

Wernicke, Martina, Corinna Hofter, Kirsten Jordan, Peter Fromberger, Peter Dechent, and Jürgen L. Müller. 2017. "Neural Correlates of Subliminally Presented Visual Sexual Stimuli." *Consciousness and Cognition* 49 (March): 35–52.

Wesling, Meg. 2022. "'Gay Genes' and the Contested Origins of Same-Sex Desire." *Feminist Studies* 48 (3): 790–805.

Wesner, Ashton. 2019. "Messing Up Mating: Queer Feminist Engagements with Animal Behavior Science." *Women's Studies* 48 (3): 309–45.

West, Suzanne L., Aimee A. D'Aloisio, Robert P. Agans, William D. Kalsbeek, Natalie N. Borisov, and John M. Thorp. 2008. "Prevalence of Low Sexual Desire and Hypoactive Sexual Desire Disorder in a Nationally Representative Sample of US Women." *Archives of Internal Medicine* 168 (13): 1441–49.

Westerlaken, Michelle. 2020. "What Is the Opposite of Speciesism? On Relational Care Ethics and Illustrating Multi-Species-Isms." *International Journal of Sociology and Social Policy* 41 (3/4): 522–40.

Weston, Kath. 1997. *Families We Choose: Lesbians, Gays, Kinship*. Columbia University Press.

Westphal, Sylvia Pagan. 2004. "Glad to Be Asexual." *New Scientist*, October 14. https://www.newscientist.com/article/dn6533-feature-glad-to-be-asexual/.

Whalen, Richard E., Frank A. Beach, and Robert E. Kuehn. 1961. "Effects of Exogenous Androgen on Sexually Responsive and Unresponsive Male Rats." *Endocrinology* 69 (2): 373–80.

Willey, Angela. 2016. *Undoing Monogamy: The Politics of Science and the Possibilities of Biology.* Duke University Press.

Willey, Angela, and Sara Giordano. 2011. "Why Do Voles Fall in Love? Sexual Dimorphism in Monogamy Gene Research." In *Gender and the Science of Difference: Cultural Politics of Contemporary Science and Medicine,* edited by Jill A. Fisher, 108–25. Rutgers University Press.

Willis, Malachi, Kristen N. Jozkowski, Wen-Juo Lo, and Stephanie A. Sanders. 2018. "Are Women's Orgasms Hindered by Phallocentric Imperatives?" *Archives of Sexual Behavior* 47 (6): 1565–76.

Wilson, Elizabeth A. 2002. "Biologically Inspired Feminism: Response to Helen Keane and Marsha Rosengarten, 'On the Biology of Sexed Subjects.'" *Australian Feminist Studies* 17 (39): 283–85.

Winer, Canton, Megan Carroll, Yuchen Yang, Katherine Linder, and Brittney Miles. 2022. "'I Didn't Know Ace Was a Thing': Bisexuality and Pansexuality as Identity Pathways in Asexual Identity Formation." *Sexualities* 27 (1–2): 267–89.

Winter-Gray, Thom, and Nikki Hayfield. 2021. "'Can I Be a Kinky Ace?' How Asexual People Negotiate Their Experiences of Kinks and Fetishes." *Psychology & Sexuality* 12 (3): 163–79.

Wise, Nan J. 2014. "Donating Orgasms to Science." *Atlantic,* January 17.

Wise, Nan J. 2022. "Learning How to Work the Sexual Brain-Body Connection." *Psychology Today* (blog), December 5. https://www.psychologytoday.com/intl/blog/why-good-sex-matters/202212/learning-how-work-the-sexual-brain-body-connection.

Wise, Nan J., Eleni Frangos, and Barry R. Komisaruk. 2017. "Brain Activity Unique to Orgasm in Women: An fMRI Analysis." *Journal of Sexual Medicine* 14 (11): 1380–91.

Wolpe, Raquel E., Kamilla Zomkowski, Fabiana P. Silva, Ana Paula A. Queiroz, and Fabiana F. Sperandio. 2017. "Prevalence of Female Sexual Dysfunction in Brazil: A Systematic Review." *European Journal of Obstetrics & Gynecology and Reproductive Biology* 211 (April): 26–32.

Wong, Day. 2014. "Asexuality in China's Sexual Revolution: Asexual Marriage as Coping Strategy." *Sexualities* 18 (1–2): 100–116.

Wong, Day, and Xu Guo. 2020. "Constructions of Asexual Identity in China: Intersections of Class, Gender, Region of Residence, and Asexuality." *Feminist Formations* 32 (3): 75–99.

Wong, Edward. 2016. "Lousy Libidos: Why Do Pandas Have So Little Sex?" *New York Times,* September 30, sec. World.

Working Group for a New View of Women's Sexual Problems. 2002. "A New View of Women's Sexual Problems." In *A New View of Women's Sexual Problems,* edited by Ellyn Kaschak and Leonore Tiefer, 1–8. Haworth.

Wright, Jessey. 2018. "The Analysis of Data and the Evidential Scope of Neuroimaging Results." *British Journal for the Philosophy of Science* 69 (4): 1179–1203.

Wyckoff, Jason. 2014. "Linking Sexism and Speciesism." *Hypatia* 29 (4): 721–37.

Wynter, Sylvia. 2003. "Unsettling the Coloniality of Being/Power/Truth/Freedom: Towards the Human, After Man, Its Overrepresentation—An Argument." *CR: The New Centennial Review* 3 (3): 257–337.

Xia, Jixing, Zhaojiang Guo, Zezhong Yang, et al. 2021. "Whitefly Hijacks a Plant Detoxification Gene That Neutralizes Plant Toxins." *Cell* 184 (7): 1693–1705.e17.

Xie, Qing-Ping, Bing-Bing Li, Wei Zhan, et al. 2021. "A Transient Hermaphroditic Stage in Early Male Gonadal Development in Little Yellow Croaker, *Larimichthys polyactis*." *Frontiers in Endocrinology* 11 (January 27): 542942.

Yule, Morag A., Lori A. Brotto, and Boris B. Gorzalka. 2014. "Biological Markers of Asexuality: Handedness, Birth Order, and Finger Length Ratios in Self-Identified Asexual Men and Women." *Archives of Sexual Behavior* 43 (2): 299–310.

Zack, Naomi. 2014. *Philosophy of Science and Race*. Routledge.

Zhang, Minming, Shaohua Hu, Lijuan Xu, et al. 2011. "Neural Circuits of Disgust Induced by Sexual Stimuli in Homosexual and Heterosexual Men: An fMRI Study." *European Journal of Radiology* 80 (2): 418–25.

Zheng, Lijun, and Yanchen Su. 2018. "Patterns of Asexuality in China: Sexual Activity, Sexual and Romantic Attraction, and Sexual Desire." *Archives of Sexual Behavior* 47 (4): 1265–76.

Zheng, Lijun, and Yanchen Su. 2022. "Sexual Minority Identity and Mental Health Among Individuals on the Asexuality Spectrum in China: A Longitudinal Study." *Archives of Sexual Behavior* 51 (7): 3627–36.

Zhiling, Huang. 2017. "63 Giant Pandas Born in Captivity in 2017." *China Daily*, November 8. https://www.chinadaily.com.cn/china/2017-11/08/content_34283946.htm.

Zucker, Kenneth J. 2015. "The DSM-5 Diagnostic Criteria for Gender Dysphoria." In *Management of Gender Dysphoria: A Multidisciplinary Approach*, edited by Carlo Trombetta, Giovanni Liguori, and Michele Bertolotto, 33–37. Springer Milan.

INDEX

ableism, 2, 24, 80, 84, 151n15; challenging, 9, 98

abstract sex, 120, 121

Abstract Sex (Parisi), 120, 177n32

Ace Community Survey, 150n1

ace-friendly, 4, 20, 102

aceness, 79, 82; animal, 97–100

acephobia, 5

ace political project, 100, 140–41, 142–43; antinaturalist intersectional, 140, 142, 144–47; intersectional/queer/feminist, 144, 147

activism: asexual, 44, 45, 79, 142–43; feminist, 45, 131, 146; LGBTQ+, 81, 131; queer, 79, 146

Addyi, 1, 43

Ågmo, Anders, 89, 166n36

Alaimo, Stacy, 96, 97

allosexuality, 2, 8, 11, 14, 60; privileging, 12

American Psychiatric Association (APA), 40, 44, 159n38

amoebas, 103, 104, 105, 170nn5–6

anaphrodism, 33

androgens, 88, 89, 90

aneroticism, 26, 32, 33

anesthesia, 28, 30, 31, 33, 155n18. *See also* sexual anesthesia

anhedonia, 33, 34

animal sexuality, 102; compulsory sexuality and, 79; defining, 3–4; human sexuality and, 83; queer/feminist critiques of, 80–84

anorgasmia, 49

antinaturalist approach, 98, 140, 142, 144, 146–47

anxiety, 45, 65, 66, 69, 145

aphrodisiacs, 26, 87, 154n9

apomixis, 108, 117

Armin, Kadji, 278n4

aromantic (aro), 5, 8

aromantic spectrum (aspec), 133, 150n8

aromanticism, romanticism and, 133

arousal, 56; emotional, 66, 67; physiological, 63, 161n16. *See also* sexual arousal

arousal disorders, 43, 44, 145, 158n33; female, 42, 71

artificial insemination, 77, 78, 176n25

asexual communities, 8, 95, 103, 133; asexual/ace persons and, 1; disabled community and, 9; exploring, 5–6; nonhuman animals and, 25, 104; online, 69; protests by, 92

asexual people, 1, 2, 9, 17, 34; amoebas and, 103, 104, 105

asexual phenomena, 18, 93–94, 166; nonhuman animals and, 86, 88, 91–92, 94; scientific research on, 16, 86–90, 92, 168n12

asexual possibilities, 3, 4–12, 22, 77, 122; analytics of, 4, 17, 18, 19, 51, 74, 75, 78, 102, 136; studies on, 51, 94–97

asexual reproduction, 16, 26, 103, 104, 125–30; attention to, 116, 134, 187; bacteria and, 19, 105–6; biological reproduction and, 174n23; birth rate for, 110; complexity of, 107, 114; compulsory sexuality and, 19; engaging in, 120, 122–23, 127; feminist studies on, 119–24; queer studies on, 119–24; scientific research on, 102, 105–11,

asexual reproduction, *continued*
112–13, 113–19, 126; sexual reproduction
and, 109, 118, 127; types of, 106, 107, 108,
112, 128
asexual spectrum (aspec), 5, 133, 140, 146
Asexual Visibility and Education Network
(AVEN), 4–5, 6, 95, 103; asexuality and,
150n6; languages of, 150n4; pandas and,
76; sexual attraction and, 150n7; survey
by, 149n4
asexuality, 10, 39, 40, 123, 133; acceptance of,
95; analytics of, 47, 109, 137; asexual phenomena and, 93–94; concept of, 6, 104;
curing, 94; deeming, 26; definitions of,
17, 86, 104, 169n1; disabled and, 99–100;
as dysfunction, 137; effeminacy and, 78;
environmental factors and, 95; gerbil,
88–89; human, 19, 86, 94, 104; identity
model of, 6, 9; marginalization of, 12;
myths about, 103; nonhuman animal, 19,
79, 86–90, 90–94, 95; panda, 76, 78, 104;
prosex biases and, 3; self-identification of,
44; as subclassification, 155n17; stigmatization of, 7, 46, 72; term, 104, 155n17,
169n1; thinking about, 16; understanding,
6, 8, 11, 58, 142
asexuality studies, 2, 4, 279n6; feminist/
queer, 6, 7, 8, 11, 13
aspec. *See* aromantic spectrum; asexual
spectrum
Atlantic, The, 49
attraction: compulsory, 8; differentiated,
150n8; mode of, 179n4; sensual, 134;
understanding, 18, 52
autism, 93
autogynephilia, 154–55n15
automonosexual, 32, 33, 154n15, 155n17
AVEN. *See* Asexual Visibility and Education
Network
Avise, John C., 123

bacteria, 114, 120, 121, 122, 130; asexual
reproduction and, 19, 105–6; donor, 128;

genetic material for, 106; lateral gene
transfer and, 127, 178n36; sex and, 125,
126, 129, 135
Bagemihl, Bruce, 101
Barad, Karen, 149n2
behavior: animal, 3–4, 80, 95; copulation-
like, 101; disarticulation of, 83;
nonreproductive, 101; observable, 86;
pair-bonding, 85; reproductive, 88, 89,
98; sex/gender specific, 95. *See also* sexual
behavior
Bell, Graham, 109
Bergler, Edmund, 156n25
Bergson, Henri, 177n32
Berridge, Kent, 65
Bianchi-Demicheli, Francesco, 162n21
biases, 139, 160n7; antiasexual, 113; prosex, 3,
114, 116
Billiard, Sylvain, 127
binary, 121; sex/gender, 54, 128; sexual/
nonsexual, 74, 79, 102, 143, 144
Biological Exuberance (Bagemihl), 101
bisexuality, 33
Bivins, Roberta, 178n36
Black women, asexual, 8, 151nn13–14,
153–54n7
Blanchard, Ray, 154–55n15
Bloch, Iwan, 33
Boast, Hannah, 98
bodymind, 99, 169n21
Boehringer Ingelheim, 43
Bonaparte, Marie, 156n24
bonding: pair-, 22, 84, 85; same-sex, 81;
social, 46, 101–2, 135, 136, 166n35
Bonduriansky, Russell, 118
Borg, Charmaine, 67
brain: gender identity and, 14; male/female,
52; organization of, 54, 95; reward-specific
areas of, 68; sexual response and, 61, 62,
64, 65, 66, 70
brain activation, 52, 65, 70, 163n26; differences in, 165n33; erotic stimuli and,
161–62n21; heterosexual male, 161n20;

imaging, 161n18; measuring, 69, 161n18, 164n29, 165n32; isolating, 53; sexual, 66, 164n27; stimuli and, 68, 69; videos/pictures and, 64, 161n21; vss and, 63–64

Brain Storm (Jordan-Young), 54

brain structure: gender identity and, 14, 54; sexuality and, 14

breeding: in captivity, 75, 77, 167n5; inbreeding and, 128

bremelanotide (Vyleesi), 43

Brotto, Lori, 14

Bühler, Mira, 161–62n21

Burke, Nathan, 118

Cacchioni, Thea, 42

care-receiving/caregiving, 146

castration, 31, 168n11

celibacy, 7, 154n8

Cera, Nicoletta, 61

Chambers, Kathleen C., 90

Charnas, Suzy McKee, 177n29

Chen, Lian, 114

Chen, Mel, 104

Chen, P., 161n18

chromosomes: homologous, 107, 110, 173n19; repairing, 111

Cikara, Mina, 62

Clark, Mertice M., 88, 89

Clarke, Esther, 102

class, 8, 23, 25, 44, 144

clitoris, 36, 156n24, 159n1

clone, 108, 116, 118, 121, 123, 171n13; adapted, 117; invertebrate, 117; term, 178n38

Colegrave, Nick, 115

colonialism, 2, 23, 24, 28, 61, 112, 113, 116, 117; racism and, 9; sexuality and, 9

competition, 111; clonal, 117; intraspecies, 112

compulsory sexuality, 4–12, 15, 16, 114, 130; analytics of, 3, 4, 11–12, 17, 18, 19, 22, 47, 51, 72–73, 74, 79, 102, 105; animal sexuality and, 79; asexual reproduction and, 19; concepts of, 131; contributing to, 46;

deviation from, 8–9; disparate activities and, 125; economic systems of oppression and, 11; focus on, 13; influence of, 46, 85; nonhuman animals and, 90–94; racial capitalism and, 11; reinforcing, 16, 105; term, 2, 7, 10, 11, 151n15; undermining, 119; voles and, 84–86

conjugation, 106, 125

Connell, R. W., 149n1

conversion therapy, 93

Cooper, Danielle, 6

copulation, 86, 87, 88, 136; advantages of, 117; orgasm and, 169n22; preventing, 27; rate of, 77; term, 129

Crow, James F., 115

Cryle, Peter: on frigidity, 24, 153n1, 153n5

Darwin, Charles, 119, 175n25

debility, 79, 93, 100

Decker, Julie Sondra, 104

degeneration, 29–30, 33, 39

de Jong, Peter J., 67

desexualization, 7, 10, 25, 142

desire, 56, 132, 162n21; disarticulation of, 83; hypersexual, 154n7; hyposexual, 16, 17; responsive, 71, 158n36; same-sex, 41; spontaneous, 71, 158n36; understanding, 18, 52. *See also* sexual desire

Deutsch, Marie, 156n24

development, 42–45, 74, 112, 175n25; asexual, 138, 139; fetal, 54, 55; of identity, 6, 150n9; individual, 69; nonsexual, 75; sexual, 75, 138, 139

developmental systems theory, 143

Diagnostic and Statistical Manual of Mental Disorders (DSM), 40, 41, 145, 159nn37–38; data in, 50; diagnostic criteria from, 43, 71; sexual desire and, 44, 45

difference studies, 58, 60, 165n33

dimorphism, sexual, 110, 114, 116, 172n16, 172n18

diploid (2n), 107, 109, 127; organisms, 108, 128, 171n12, 178n35; zygotes, 127, 171n12 disability, 8, 20, 23, 25, 41; asexuality and, 99–100; pride model of, 100; racism and, 99; reducing/eliminating, 93; sexual violence and, 151n12

disability studies, 9, 99

discrimination: compulsory sexuality and, 8; in health care, 5; racial, 175n25

disgust responses, 67, 68, 69, 164n28

Disorders of Sexual Desire and Other New Concepts and Techniques in Sex Therapy (Kaplan), 38–39, 158n33

diversity, 96, 114–18, 144, 145; genetic, 106; and reproduction, 116, 121, 125, 128; sexual, 94, 121, 124, 125, 132, 147; support for, 20

DNA, 106, 170n4, 171n12, 176n28; mitochondrial, 123; sexual reproduction and, 111

domination, 80, 136, 141, 166n35; genital contact and, 102

Door into the Ocean, A (Slonczewski), 176n29

drive model, 43, 71, 72, 73, 136; dropping, 58

DSM. See *Diagnostic and Statistical Manual of Mental Disorders*

Duchesne, Annie, 55

duds, 88, 90; term, 18, 74, 79

Dussauge, Isabelle, 55, 69, 160n12, 163n26; on sexuality, 61, 65; on sexual response system, 65–66; on sexual stimuli, 64, 162n22

dyspareunia, 40

ecology, study of, 116, 117

education, 13, 16, 52; sex, 9, 24, 44

ejaculation, 33, 40, 77, 87, 90, 101

Ellis, Havelock, 33, 34, 155n18

endocrine disruptors, 97

endocrine function, 88

endosymbiosis, 126, 170n4, 177n31

environmental cues, 71, 78, 95

environmental toxins, 97, 98; nonhuman animals and, 100; queer animals and, 99

Epistemology of the Closet, The (Sedgwick), 139

erectile dysfunction (ED), 6, 21, 39, 42, 93; focusing on, 59, 61; psychogenic, 83, 84

eros, 141, 142–43, 179n9

erotic, the, 63, 140–41, 163n26, 164n27, 165n32; and asexuality, 6, 142, 144, 146–47, 151n16; brain response to, 68; negative-context, 165n31; sovereign, 11; term, 141, 144

estradiol, 87

estrogen, 90, 67n11

ethics, 29, 30, 92, 94

eugenics, 23, 28, 93, 99, 112; positive, 175–76n25; racism and, 176–77n30; rejection of, 113

Eugenics Congress, 175n25

eukaryotes, 106, 107, 109, 110, 123, 127; sexual reproduction and, 118

evolution, 69, 107, 109, 119; sex and, 136; study of, 116, 117

evolutionary biology, 109, 112, 113, 115, 124, 175n25

experiences: asexual, 103, 154n8; emotional, 57; gendered, 160n10; lesbian, 2; nonsexual, 149n3; sexual, 3, 57, 157n28, 166n36; trans, 15, 155n15

extinction, 75, 109, 114, 177n33

Farah, Martha J., 52, 53

FDA (Food and Drug Administration), 1, 44

Feder, Harvey H., 88

Female Man, The (Russ), 176n29

Female Sexual Distress Scale, 43

female sexual dysfunction (FSD), 42, 44

Female Sexual Function Index, 43

feminist studies, 16, 53, 54–57, 80–81, 83; asexual/sexual reproduction and, 119–24; binary sex/gender system and, 54; neuroscience research and, 55; sex drive and, 133, 134, 139; sexual desire and, 82

feminists, 158n32; queer/sex-positive, 144; sexual desire and, 132

Feng, Chunliang, 165n32

fertilization, 108–9, 127, 135, 171n12, 172n14; double, 170n10; in vitro, 174n23

Fisher, Ronald A., 113, 175n25
Fisher-Muller hypothesis, 113, 173n20, 175n25
flibanserin, 43–44
fMRI. *See* functional magnetic resonance imaging
Foucault, Michel, 126, 153n2; sexual desire and, 132, 133, 178n1
Freed, S. Charles, 156n25
Freud, Sigmund, 155n21; libido and, 35, 57–58, 72; life drive and, 142; sexual desire and, 36, 57–58, 71; sexual drive and, 34–36
frigidity, 17, 22, 24, 31, 33; cases of, 26, 27, 155n18; definitions of, 36–38; diagnosis of, 156n24; female, 36, 156n23, 157n31; term, 156n26, 158n32; understandings of, 156n24, 158n32
Frigidity (Cryle and Moore), 153n5
frogs, 76, 97; gay, 98; sexually inactive, 100
Fuechtner, Veronika, 3, 153n3
functional magnetic resonance imaging (fMRI), 1, 14, 52, 159nn1–2; studies, 48, 49, 56
fungi, 106, 107, 114, 170n8, 179n7; asexual reproduction of, 123, 127

Galef, Bennett G., 88, 89
gametes, 108, 126, 127, 129, 173n18; formation of, 112, 171n14; fusion of, 118, 121; male/female, 112, 172n16; mobile/immobile, 13; sexual reproduction and, 107
Garland-Thomson, Rosemarie, 79, 93
gender, 20, 25, 28, 126, 131; Black/Indigenous systems of, 10; as cultural construction, 96–97; discrepancy, 41; essentialist thinking about, 15; expression, 60; more-than-human, 96; neuroscience research on, 54–57; nonconsensual altering of, 99; nonnormative, 22, 23; norms about, 10–11; sex and, 122, 160n8; sexuality and, 9, 10, 56, 96; social concepts of, 8, 15, 23
gender dysphoria, 159n37
gender identity disorder, 24, 155n17, 159n37

gender nonconformity, 9, 14
genes, 14, 166n36, 168n11; deterministic view of, 176n25; exchanging, 122; repairing, 110, 111; transfer of, 169n3, 170n4
Genetical Theory of Natural Selection, The (Fisher), 175n25
genetic engineering, 126, 128
genetic material, 106; combining, 129, 136; passing, 89; reproduction and, 127, 128; sex and, 126–27
genetic recombination, 126, 127, 136, 173n22
genetic shuffling, 111, 117
genetic variation, 109, 113, 175n25
genital contact, 133, 167n9; dominance and, 102; same-sex, 101–2
genital stimulation, 14, 73, 137, 158n32, 159n3; partnered, 101
genomes, 106, 111, 172n16, 178n35; bacterial, 169n3; host, 170n4; prokaryotes and, 110
genomic combinations, 118, 171n14
genotypes, 172n16, 173–74n22, 175n25
Georgiadis, Janniko R., 67, 163n24
gerbils, sexual/reproductive behavior of, 88–89
germ plasm, theory of, 175n25
Ghost Stories for Darwin (Subramaniam), 112
giant pandas. *See* pandas
Gillis-Buck, Mae, 171n14
Gilman, Charlotte Perkins, 119
Ginzberg, Ruth, 179n9
Girl One (Murphy), 177n29
Gola, Mateuz, 162n23
Goldhammer, Denisa, 57
gonochorism, 112, 129, 172–73n18
Gramsci, Antonio, 149n1
gray aromantic, term, 133
gray asexual, term, 133
Grosz, Elizabeth, 119
guinea pigs, 88
gynogenesis, 171n12

Hammond, William Alexander, 30

haploid (n), 108, 109; cells, 170n5, 170n8, 176n28; eggs, 171n12

Haraway, Donna, 96

Harding, Cheryl F., 88

Harshaw, Christopher, 138

Hartfield, Matthew, 173n20

"Haven for the Human Amoeba" (Yahoo group), 103

Haynes, Douglas E., 153n3

Heintzel, Alexander, 164n29

Helm, Bennett, 179n9

hermaphroditism, 22, 108, 170n5, 173n18; self-fertilizing and, 127; sexual selection in, 175n24

heterosexism, 2, 60, 84, 151n15, 179n8; challenging, 98

heterosexual-homosexual scale, 37, 157n27

heterosexuality, 3, 11, 33, 40, 55, 96; advantages of, 116, 117; compulsory, 2, 130; imposition of, 178n36; as subclassification, 155n17

heterosexual reproduction, 120, 125, 174n23; decentering, 123; importance of, 121

Hiebert, Laurel S., 116

Hill, Geoffrey E., 173n20

Hill-Robertson interference, 173n20

Hird, Myra, 15, 96, 122, 124, 125; asexual phenomena and, 97; asexual reproduction and, 120; microontology and, 121; on sexual difference, 123

Hirschfeld, Magnus, 24, 26, 28, 45, 155n15; anesthesia/frigidity and, 155n18; on asexuality, 155n12; heterosexual classifications by, 155n16; sexology and, 32, 33; on sexual disinterest, 33; theory of intermediaries and, 179n4

History of Sexuality, The (Foucault), 132, 153n2

Hitschmann, Eduard, 156n25

homophobia, 7

homosexual, as category or classification, 40, 55, 155n17

homosexuality, 22, 28, 30, 33, 40; curing, 15;

decriminalization of, 155n10; ego-dystonic, 159n38; existence of, 93; hypoactive desire and, 41; as natural, 81; nonhuman animal, 82, 86, 88–89, 167n9; nonpathological view of, 35

hormone replacement therapy (HRT), 154n13

hormones, 54, 154n13, 166n36, 168n11; administration of, 134; exposure to, 82; gonadal, 87; prenatal, 14; sexual, 14, 163n25

HSDD. See hypoactive sexual desire disorder

Human Sexual Inadequacy (Masters and Johnson), 37–38

Human Sexual Response (Masters and Johnson), 37

hyperesthesia, 28

hypersexuality, 9, 121, 151nn13–14, 154n7; diagnosing, 160n13; shared, 10; studies of, 23, 59

hyphedonia, 33

hypoactive desire, 39, 41

hypoactive sexual desire disorder (HSDD), 10, 17, 22, 26, 38–40; brain and, 58, 59; definitions of, 31–32; development of, 42–45, 72, 93; diagnosing, 41; scientific research into, 1, 87; women and, 88, 91, 168n17

hyposexuality, 10, 16, 17, 25

Ideal Marriage (van de Velde), 156n22

identity, 143, 150n6, 153n2, 165n33; asexual, 4–12, 16, 17, 18, 25, 27, 54, 92, 95, 103, 104, 130, 142; disability, 100; disarticulation of, 83; gay, 82; gender, 54, 95, 150n11, 155n16, 179n4; lesbian, 82; self-perceived, 139; sexual, 4, 7, 9, 11, 20, 23, 24, 82, 91, 103, 104, 131, 150n9, 165n33, 179n4; transgender, 55, 150n11

images: bodily/nonbodily, 165nn30–31; brain, 52; emotional, 67, 165n32; erotic, 63, 161n16, 163n25, 164nn27–28; sexual, 62, 67, 68, 165n30, 165n32; sports, 164n30 impotence, 26, 32, 36, 40, 152n1; cerebral, 33; cures for, 154n9, 154n13; redefining, 42

inbreeding, breeding and, 128

incentive motivation model, 43, 71

inequality, 80; gender, 54, 61, 126, 174n23; sexual, 61

inhibited sexual desire (ISD), 21, 22, 24, 26, 36; anxiety and, 45; focus on, 38, 40, 42

instinct, 34, 35, 36, 73; primal, 155n21; sex, 155n21; sexual, 28, 30, 31, 154–55n11

intercourse, 26, 31, 33, 157n30; heterosexual, 50, 158n32; nonenjoyment of, 152n1; partnered, 37; term, 129; vaginal, 36, 157n31

International Affective Picture System (IAPS), 165n31

International Union for Conservation of Nature, 78

intersectionality, 8–9, 15, 55, 60, 80; antinaturalist approach to, 140, 142, 147

intersexuality, 22, 60, 97

intimacy, 129, 130, 141; need for, 7; physical/emotional, 146; sexual, 3

Intrinsa, 43

Invisible Orientation (Decker), 103–4

involuntary celibacy (incels), 7

Irigaray, Luce, 119

Irvine, Janice M., 24, 38, 156n26

Japanese macaques, 83, 92, 101, 169n22

Johnson, Virginia, 36, 38, 157nn30–31, 158n33; research by, 37, 50, 51, 58; on sexual desire, 37; on sexual response cycle, 40

Jones, Ryan M., 153n3

Jordan-Young, Rebecca, 54

justice: crip, 120; environmental, 99; gender, 112; racial, 120; reproductive, 99; social, 124; trans, 120

Kaiser, Anelis, 55, 160n12

Kaplan, Helen Singer, 21, 24, 158n33; on desire phase disorders, 38–39; on homosexuality, 41; on HSDD, 39; on ISD, 22, 26,

45; on psychosexual disorder, 40; on sex therapy, 38; on sexual desire, 58, 71

Katz, Larry S., 89

Keightley, Peter D., 173n20

Kenney, Theresa N., 165n34

Kinsey, Alfred, 36, 37; and heterosexual-homosexual scale, 156n27; on sexual activity, 156n28; on sexual capacity, 156n26; on sexual dysfunction, 157n31

kinship, 147, 174n23

Kirksey, Eben, 122, 125, 177n33

Klein, Sanja, 163n24

kleptogenesis, 117, 171n12

Koedt, Anne, 158n32

Komisaruk, Barry, 48, 159n5; study by, 49, 50–51, 159n1, 159nn3–4

Kondrashov, Alexey, 111

Kondrashov's hatchet, 173n20

Kono, Tomohiro, 171–72n14

Krafft-Ebing, Richard von, 24, 45; on contrary sexual instinct, 154n11; sexology and, 28, 34; on sexual anesthesia, 28, 29, 30, 31, 155n18; on sexual desire, 31; on sexual disinterest, 33

Kragel, Philip A., 68

Kriegeskorte, Nikolaus, 160n7

Kringelbach, Morten L., 65

Kroger, William, 156n25

Laan, Ellen, 166n36

Laplanche, Jean, 34, 35, 155n21

lateral gene transfer, 114, 117, 169n3, 176n28; bacteria and, 127, 178n36; evidence of, 106; importance of, 107; types of, 125–26

Lehtonen, Jussi, 172n16

Leif, Harold, 38, 40, 58, 158nn33–34

lesbian bed death, 41

lesbian feminists/separatists, parthenogenesis and, 120

libido, 32, 150n7, 156n26, 158n33; Freud on, 35, 57–58, 72; improving, 154n13; lack of, 25; variation in, 38 lifeforce, 141, 142

localization, criticism of, 53

Lorde, Audre, 179n9; ace political project and, 140–42; on the erotic, 141, 142, 143, 144; on pleasure, 163n26

lordosis behavior, 82, 168n11

Lucretius, 177n32

macaques, Japanese, 83, 92, 101, 169n22

marginalization, 2, 10–11, 98, 99; asexuality and, 72; experiencing, 5; nonsexuality and, 7, 12, 15; privilege and, 8

Margolin, Leslie, 45

Margulis, Lynn, 116, 118, 120; on lateral gene transfer, 176n28; on sexual reproduction, 124; on symbiogenesis, 177n30

marriage, 11, 21, 146; manuals for, 36, 156n23; reproduction and, 27; same-sex, 81; shoring up, 157n30

Married Love (Stopes), 156n22

Martens, Koen, 129

Martynova, Olga V., 161n16

masculinity, hegemonic, 8, 9, 90–91, 149n1

masochism, 22, 28

Masters, William, 36, 38, 157nn30–31, 158n33; research by, 37, 50, 51; on sexual activity, 58; on sexual desire, 37; on sexual response cycle, 40

masturbation, 22, 35, 37, 50, 157n30; non-human animal, 179n5; solo, 121

mating, 129, 135, 136, 170n7; heterosexual, 120, 125; tests, 87

Maynard Smith, John, 110, 113, 176n25

McCabe, Marita, 57

Meana, Marta, 71

Medical Entanglements (Gupta), 42, 45, 168n16

medical health, 22, 23, 25, 26, 27; intervention in, 45, 79; sexual desire and, 134

meiosis, 110, 121, 128, 170n6, 171n12; cost of, 172n16; evolution of, 118; genetic material and, 107, 111, 126; modified, 117; parthe-nogenesis and, 127; single-celled gametes and, 108

Meirmans, Stephanie, 175n25

mental health, 9, 35, 32, 36, 134; discourse for, 23, 25; treatments for, 33, 41

Michod, Richard E., 114, 178n35

microbiology, 120

microorganisms, 104, 121, 122

mind/body dualism, Cartesian, 53

mitochondria, 121, 123, 170n4, 173n20, 177n31, 177n34

mitosis, 107, 117, 121, 123

mixis, 14, 118

Moccia, Lorenzo, 72

Moll, Albert, 33

Monera (kingdom), 121

monogamy, 11, 85, 94, 96; compulsory, 18, 84

Moore, Alison O., 24, 153n5

morphology, 112, 129, 171n13, 176n28

Morton, Timothy, 123

mosaicism, 54, 74

Motherlines (Charnas), 176n29

motivational states, 65, 73, 137

Muller, Herman J., 113, 175n25

Muller's ratchet, 173n20, 175n25

multicellular organisms, 13, 124, 177n35

Murphy, Sara Flannery, 177n29

mutations, 116, 173n20, 178n35; genetic, 106, 175n25; sexual reproduction and, 111

naloxone, 87, 93

National Zoological Park, 77

naturalness, 81, 93, 94, 96

natural selection, 113, 175n25

nature, queerness of, 19, 105

neoliberal capitalism, 11, 132, 163n26

neurasthenia, sexual, 35

neurocentrism, 53

neurofeedback therapy, 49–50

neurofeminism, 54, 55

neuroimaging, 3, 71, 160n7, 161n18, 164n27;

analysis of, 17, 64; growth of, 52; sex/gender, 55, 160n9

neuroimaging research, 16, 17, 51, 56, 160n9; sex/gender, 52–53, 54, 60; using, 61, 62, 71–72, 73, 74

neuroimaging studies, 47, 51, 53, 54, 56; analysis of, 17–18, 59, 60, 61, 67; designing, 69; "double" comparison and, 164n27; responses and, 64

neuropharmacological treatments, 59

neurophenomenological model, 72

neuroscience, 48, 72, 136, 142, 160–61n15; critical engagements, 52–54; feminist engagements, 52, 54–57; growth of, 54, 60; sexism in, 160n10

Neuroscience (journal), 1

neurosis, 35

neurotransmitters, 166n36, 168n11

New Sex Therapy, The (Kaplan), 38, 155n20

New View Campaign, 44

New York Times, 75

noncopulating male rats, 1, 91; sexual activity and, 87, 93, 94; testosterone and, 86, 92

nonhuman animals: ace people and, 82; altering lives of, 99; asexuality and, 19, 78, 79, 86–90, 90–94, 95; compulsory sexuality and, 90–94; environmental toxins and, 100; homosexuality in, 81, 83, 86, 88–89; human animals and, 79–80; queer, 120; sexual behavior and, 18, 80, 81, 84, 91–92, 100–101, 165n36, 168n19; sexuality and, 18, 79, 82, 91, 92, 95–96, 97, 100–102, 134, 161n18

nonsexual, 8, 35, 66, 67–68; sexual and, 4, 16, 69, 73, 144, 146; term, 131

nonsexuality, 11, 15, 73, 144, 152n7; conceptualizing, 46; as dysfunction, 137; marginalization of, 7; nonreproductivity and, 78; pathologization of, 72; scientific research on, 79; sexual desire and, 139; social acceptance of, 46; understanding, 74

Not Gay (Ward), 133

nymphomania, 22, 154n7

Oosterhuis, Harry, 30

oppression: gender-based, 112; sexual, 140; systems of, 8, 11, 32, 62, 141

O'Reilly, Zoe, 103

organisms, 34–35, 105, 167n9; multicellular, 123; nonhuman, 122; transgenic, 128

orgasm, 37, 64, 159n3, 159n6; achieving, 50; asexual studies discussion of, 160n6; copulation and, 169n22; experiencing, 36, 50, 158n32, 169n22; female, 160n6, 179n6; fMRI studies of, 48, 49; neuroimaging studies of, 61; problems with, 21, 39, 152n1, 158nn32–33; sex based on, 179n6; vaginal intercourse and, 157n31

orientation: asexual, 14, 92, 95, 169n21; disarticulation of, 83; gender, 93, 95; homosexual, 83. *See also* sexual orientation

Origins of Sociable Life, The (Hird), 121

Orive, Maria E., 126, 129

Ortigue, Stephanie, 162n21

Otto, Sarah, 115, 123, 172n17

outcrossing, 127, 170n8, 178n35

Owen, Ianna Hawkins, 141, 142, 151n14

Palmer, Seth, 34

pandas: asexuality of, 76, 78, 104; captive breeding and, 75, 77–78, 167n5; sex drive of, 75; sexuality of, 75–76, 77, 78, 90, 100; videos for, 166n1

parasexuality, 107, 127, 171n12

parasites, 111, 115, 173n20, 173–74n22

Paredes, Raúl G., 86, 87, 93

parental investment, 13–14, 96, 112

parenting, genetic material and, 89

Parisi, Luciana, 120, 121, 124, 125, 177n32

parthenogenesis, 119, 121, 127, 171–72n14, 172–73n18; all-female communities and, 120; meiotic, 128; mitotic/apomictic, 128; reproduction by, 108, 109, 115, 117, 118, 171n12

pathologies, 2, 5, 18, 38, 58–59; gender, 3; sexual, 22, 23, 24–25, 43, 46, 59, 62, 74
patriarchy, 2, 121, 149n1, 151n15, 169n2
pedophilia, 59, 160n13, 163n24
penguins, same-sex pairs of, 81
people of color, 80, 81, 98, 152n19, 174n23
Persson, Sofia, 160n10
Pfaus, James G., 56–57, 83
Pfizer Pharmaceuticals, 59
pheromones, 63, 161n19
Philip, Marion, 27
Phlegmariurus saururus, 87, 93
Phoenix, Charles H., 90
physiology, 3, 26, 37, 42, 63, 129
Pincus, Gregory, 171n14
platonic, term, 146
pleasure, 15, 45–46, 72, 100, 163n26; asexual, 142; diversity of, 144, 145; experiencing, 102, 126, 129, 130, 134, 144, 145, 146, 166n35; inhibition, 39; nonsexual, 143, 144; organization of, 138; orgasmic, 121; relationships and, 143; sexual, 3, 18, 36, 47, 64, 73, 102, 135, 136, 138, 142, 143, 144
pleasure/reward system, 73, 165n35
politics, 23, 122, 151n14, 176n25
Pollock, Anne, 100
polyamory, 122, 146
polykaryotic, 127, 170n8
Pontalis, J. B., 34, 35, 155n21
pornography, 163n24; brain and, 63–64; panda, 75, 76
Portillo, Wendy, 86, 87, 93
Pownall, Madeleine, 160n10
Pride Parades, 5, 150n5
Princeton Guide to Evolution, The, 178n37
problematic pornography use (PPU), 163n24
progesterone, 168n11
prokaryotes, 105, 110, 177n34
protists, 107, 114, 170n6
Przybylo, Ela, 6, 7–8, 141, 142
psychological research, 27, 53, 150n9, 160n10
psychology, 26, 35, 41, 42, 112
Psychology Today, 48

Psychopathia Sexualis (Krafft-Ebing), 28, 29, 30
psychopharmaceuticals, 3
psychosexual therapy, 21
psychotherapy, 32
Puar, Jasbir, 79

queer, 99, 100, 122
queer animals, 96; environmental toxins and, 99; feminist/queer studies on, 120
queer engagements, 52, 96, 105, 122
queerness, 79, 105; animal, 97–100; environmental toxins and, 100
queerplatonic, term, 146
queer possibilities, 77
queer studies, 82, 83, 112, 119–24, 139; neuroscience research and, 55, 60; nonhuman sexuality and, 95; sex drive and, 131, 133, 134; sexual orientation and, 80–81
queer theory, 139, 140, 141, 174n23, 178n2

rabbits, sexual activity of, 89
race, 44, 56, 62, 63, 131; sexuality and, 8, 9, 10, 20, 23, 25
racial hygiene, 3, 22, 34, 134
racism, 24, 60, 61, 84, 112; anti-Black, 2; challenging, 98; colonialism and, 9; disabilities and, 99; eugenics and, 176–77n30; institutional, 23; scientific, 34
Rado, Sandor, 158n33
rats: female orgasm of, 169n22; penile erection and, 83; sexual activity and, 86–87; studying, 83–84. *See also* noncopulating male rats
recombination, 107, 128, 172n16, 173n19, 178n37; genetic, 106, 126, 127, 128, 136, 173n22; mitotic, 117
Redfield, Rosemary J., 169n3
Redouté, Jérôme, 66, 67, 161n21
Red Queen hypothesis, 111, 115, 173n20
relationality, 45, 46, 134, 135, 142; sexuality and, 133
relationships, 143, 144; cognitive, 145;

different-sex, 81; emotional, 145; human/nonhuman, 167n7; intimate, 46; nonsexual/nonromantic, 146; platonic, 146; romantic, 146; same-sex, 81; sexual, 86, 146; twoness requirement for, 85

repair hypothesis, 110, 111

repression, 3, 32, 35, 72; sexual, 34, 58; societal, 156n23

reproduction, 27, 29, 114, 134, 136; bacterial, 121; connections to, 137; diversity of, 121, 128; failure at, 78; forms of, 120, 124; nonhuman mode of, 98, 103; nonsexual, 17, 123; options for, 99; parasexual, 170n12; patriarchal understandings of, 119; promoting, 45; sex and, 19, 105, 125; types of, 128, 129, 130; vegetative, 208. *See also* asexual reproduction; heterosexual reproduction; sexual reproduction

reproductive labor, racial capitalism and, 11

Resenbrink, Greta, 119

rhesus monkeys, 18, 90, 92

Rice, William R., 116

Rich, Adrienne, 3, 7

Right to Maim (Puar), 100

Rippon, Gina, 160nn9-10

Robinson, Paul A., 157n30

romance, 69, 77

romantic attraction, 8, 133, 134; sexual attraction and, 6, 150n8

romanticism, aromanticism and, 133

Roselli, Charles E., 90

Rosenberg, Gabriel, 166n4

Roth, Lateefah, 121

Roy, Deboleena, 81

Rubin, Gaye S., 139, 143

Ruesink, Gerben B., 163n24

Sabatinelli, Dean, 164n29

sadism, 22, 28, 164n28

Sagan, Dorian, 176n28

Sandford, Stela, 177n32

Sawicki, Jana, 132, 133

Schiebinger, Londa, 125

Schön, Isa, 129

science, 14, 104, 114; applied, 46–47; hegemonic, 2, 3, 12, 13, 18–19; political use of, 82; racial, 28; sexual, 15, 16, 25, 153n4

Scientific American, 75

scientific research, 118, 131, 134, 137, 139; basic, 17, 47, 57, 74; critiquing, 95, 112; drive model and, 58; feminist/queer, 80–81; hegemonic, 2, 13, 135; production of, 12–13

Scott, Joan, 143

Sedgwick, Eve Kosofsky, 169n20; intimacy and, 146; queer theory and, 139; sexuality and, 140, 143

selfing, 108, 127, 128, 170nn4-5; outcrossing vs., 170n8

Seo, Younghee, 163n25

serotonin, 43

sex: bacterial, 120, 121, 125, 126, 129, 135; categories of, 105, 135, 136; cytoplasmic, 121; engagement in, 27, 147; evolution of, 73, 104, 105, 115; gender and, 122, 160n8; genetic material and, 126–27; genital, 121; microbial, 120; nonsex and, 69; physiology of, 37; reproduction and, 19, 101, 105, 125; scientific research on, 54–57, 113–19; social concepts of, 4, 15; term, 116, 126, 129, 130

sex addiction, 22, 59

sex drive, 34–36, 72, 104; lack of, 32; perverse expression of, 28; strength of, 39–40; variation in, 38

sex/gender differences, 55, 160n9, 160-61n15, 164n27; measuring, 160n10; term, 160n8

sexism, 60, 62, 98

sex life, 29, 42, 145, 159n4

sexology, 25, 36, 46, 59, 71; American, 34, 153n1; diagnostic categories in, 24; eugenicist, 98; European, 27–34; as global project, 153n3; history of, 38, 55; interpretation of, 153n1

sex-positive feminist theory, 178n2

sex-positive queer theory, 178n2

sex therapy, 24, 38, 41

sex toys, 11, 101

sexual activity, 3, 7, 24, 27, 35; antinaturalist approach to, 140; brain and, 63–64; category of, 18, 74, 79, 100, 101, 129, 131, 132, 135, 139; concept of, 14, 18–19; copulation as, 136; denaturalizing, 135–39; engaging in, 86–87, 89, 100, 116, 125, 142, 157n28, 169n1; essentializing, 132–34, 135; inducing, 71, 93, 94; low levels of, 41; measures of, 89; physiological changes during, 58; pursuing/initiating, 80; same-sex, 82, 97, 133; scientific research on, 74; thinking about, 137

sexual anesthesia, 17, 22, 27–34, 45, 155n18, 155n20; congenital, 29; diagnosis of, 30

sexual arousal, 18, 54, 66, 69, 101; brain and, 61; concept of, 47; disorder, 38; experiencing, 155n15; male, 83; measuring, 162n24; models of, 56–58, 71–72; neuroimaging research on, 61, 70, 71–72, 137; sexual desire and, 57, 73; studies on, 47, 51, 64–65

sexual attraction, 22, 40–41, 63, 133, 134; ace-positive model of, 56; defining, 138; developmental thinking about, 72–74; experiencing little/no, 2, 4, 5, 8, 16, 17, 86; privileging, 11; romantic attraction and, 5, 150n8; thingification of, 133

sexual behavior, 4, 23, 88, 129, 134; category of, 74, 79; coercive, 95, 168–69n19; defining, 100–102, 167n9; hormones and, 14; inducing, 87, 94; lack of, 16–17, 99; levels of, 89, 90; male-typical/female-typical, 82, 92; nonhuman animal, 18, 80, 81, 84, 91–92, 100–101, 165n36, 168n19; recognizing, 136; same-sex, 82, 97, 98–99

sexual desire, 3, 7, 11, 17–18, 22–26; antinaturalist approach to, 142; categories of, 131, 132, 135, 136, 139; concept of, 17, 22, 31, 47, 54, 56–57, 135, 179n7; denaturalizing, 132, 133, 135–39; describing, 39; disorders of, 21, 25, 40, 43, 44, 158n33; drive model of, 42–43, 72–73;

134, 136; emotional/motivational states and, 51; essentializing, 132, 134, 135, 278; female, 92, 132–33; lack of, 25, 40, 43, 44; low, 1–2, 20, 45, 46; male, 132–33; medical/mental health and, 134; models of, 56–58, 71–72, 73, 137, 143, 168n11; neuroimaging research on, 16, 56, 57, 70, 71–72; nonsexuality and, 139; responsive, 57, 70; sexual arousal and, 57, 73; studies of, 57, 64–65; term, 64, 150n7; testosterone and, 92; understanding of, 3, 18, 19, 47, 131. *See also* inhibited sexual desire

sexual deviance, 9, 10, 23, 24

sexual difference, 66, 96, 113, 119, 123, 129

sexual disinterest, 10, 28, 29, 33, 91; categories of, 36–37; diagnoses for, 22–26, 36, 152n1, 154n7; European ideas about, pre-late nineteenth century, 26–27; medical approaches to, 42; as mental illness, 40; pathologization of, 25, 26, 33, 35, 42–45, 46; sexual disorder and, 33; understanding of, 17; white-settler societies and, 26–27

sexual disorders, 70, 154n13; drug treatments for, 83; focus on, 36–38

sexual dysfunction, 21, 37, 38, 42, 50; clinical significance criterion for, 40

sexual health, 9, 24, 43

Sexual Impotence in the Male and Female (Hammond), 30

sexual incentive motivation (SIM), 167n11

sexual interest, 10, 45, 104, 155n18; medicalization of, 44; nonhuman animal, 91–92

sexuality, 28, 37, 39, 42, 54, 61; active, 91; asexuality and, 133; Asian, 78, 151n14; colonialism and, 9; conceptions of, 15, 71, 138; denaturalizing, 136; disidentification with, 7, 9; drive model of, 42–43, 58, 71, 72, 73, 136; dual control model of, 161n17; female, 14, 44, 60, 71, 80, 157n30; feminist analysis of, 120; gender and, 9, 10, 96; male, 14, 80, 157n30; minority, 10, 94–95; models of, 51, 61, 96; neuroimaging research on, 72–73, 75; neuroscience research on, 54–57; nonhuman animal, 18, 79, 91, 92, 95–96, 97, 100–102, 134; nonnormative, 22, 23, 24, 62, 94, 166n3, 167n8; physiological aspects of, 50; queer, 6, 79, 100, 120; race and, 8, 9, 10, 20, 23, 25; reproductive, 10, 19, 101, 125; scientific research on, 3, 12–16, 19, 24, 49, 54, 59, 79, 93, 131, 135, 139, 140; sex/gender differences in, 13–14, 60; understanding of, 56, 74; vaginal, 36; white settler systems of, 9, 10, 11. *See also* compulsory sexuality

sexual minorities, 24, 169n20, 174n23

sexual motivation, 28, 43, 58, 65, 97; concept of, 94; decrease in, 32; lack of, 86; motivational states and, 137; nonhuman animal, 100, 165n36; states/feelings and, 73; term, 64

sexual nonattraction, developmental thinking about, 72–74

sexual norms, 7, 10–11, 23, 50

sexual orientation, 39, 54, 63, 82, 83, 94, 131; binary, 55; essentialist thinking about, 15; gender/race and, 56; genetic component of, 14; minority, 93, 95; neuroimaging studies of, 61; normal/natural, 7; sex/gender and, 60; stimuli and, 68

Sexualpathologie (Hirschfeld), 32, 33

sexual reproduction, 98, 106, 128–29, 134, 136; advantages of, 115, 116, 173n21; asexual reproduction and, 109, 118, 127; bias for, 19, 102, 105, 113–19; complexity of, 107, 114; compulsory sexuality and, 19; cost of, 110, 172n16; eugenics and, 113; evolution of, 104, 112; explanations for, 112, 113; feminist studies on, 119–24; genetic material and, 127, 128; parthenogenic reproduction and, 118; queer studies on, 119–24; scientific research on, 112–13, 119; variation and, 116, 175n25

sexual response, 50, 58, 63, 68; brain and, 59–60, 61, 62, 64, 65, 66, 69, 70, 162n24, 164n27; models of, 56–58, 61, 65, 71–72,

137; neuroimaging research on, 51, 56, 60, 61, 62, 64, 70, 71–72; stimuli for, 137; studies of, 64–65, 67; system, 3, 16–17, 59–60, 61, 70, 72, 73, 137, 162n24, 164n27

sexual selection, 60, 95, 112, 119, 168n18

sheep: FOR/MOR/NOR orientation of, 90–91; sexual/reproductive behavior in, 89–90

sildenafil (Viagra), 42, 75, 83

slavery, 9, 113, 151n13

Slonczewski, Joan, 177n29

Snoeren, Eelke M. S., 87

social categories, 25, 122, 130, 140

social meanings, 73, 137

social monogamy, 84, 85

social norms, 13, 55, 73–74, 138, 163n26

social regulation, 7, 23

Society for Neuroscience, 48

Soucany, Jean Joseph, 27

speciesism, 80, 152n19

sperm, 97, 171n12, 176n25

Spillers, Hortense, 151n13, 154n7

Spinoza, Baruch, 177n32

Sprout Pharmaceuticals, 43

Spurgas, Alyson K., 42, 60, 71

Stark, Rudolf, 164n28

Stein, Ed, 82

Stekel, Wilhelm, 156n24

stereotypes, 14; Asian/Asian American, 153n6; conflicting, 151n13; disabled, 153n6; gender, 112, 160n9; hypersexualized, 151n14; pedophile, 160n14; racialized sexual, 10, 78

sterilization, 30; eugenic, 24, 175n25

stimuli: bodily, 73, 137; brain activation and, 69; environmental, 138; erotic, 14, 69, 70, 161–62n21, 162n24, 165n31; genital, 49, 136, 159n1; heterosexual/homosexual, 156n27; incongruent, 68; neutral, 69, 164n27; nonsexual, 64–65, 66, 164–65n30; preferred/nonpreferred, 68; romantic, 69; sexual, 61–66, 68, 73, 137, 138; types of, 67–68. *See also* visual sexual stimuli

Stoléru, Serge, 72, 162–63n24

Stone, Aaron J., 153n1

Stopes, Marie, 156n22

studs, 88

Sturgeon, Noel, 81

sublimation, 35–36, 58

Subramaniam, Banu, 112

subtraction method, 53, 70, 162n24

suicide, 5; race, 46, 98

Sukel, Kayt, 49, 50, 51, 159n2

surgery, sex-/gender-affirming, 9

Sylva, David, 68

Talia, 104

TallBear, Kim, 10, 11, 151n15

technology, 3, 13; brain scanning, 58–59; imaging, 134; neuroimaging, 59, 62–63, 70; reproductive, 174n23

Terry, Jennifer, 82

testosterone, 87, 88, 90, 93, 97; blood concentration of, 89; effectiveness of, 43; female sexual desire and, 92; noncopulating male animals and, 92; synthetic, 154n13

testosterone propionate (TP), 89, 90

Tiefer, Leonore, 38, 42

transgender, 9, 22, 150n11, 179n4; authenticity of, 96; sexuality of, 155n17

transsexuality, 22, 97

transvestism, 33, 155n15

trauma, 93, 166n3, 167n8; gendered, 42; sexual, 42, 95

Trieb, 34

Trujillo, Anelis Kaiser, 55

Uddin, Lisa, 77

UNESCO, 175n25

US Food and Drug Administration (FDA), 1, 43

vagina, 36, 101, 159n1; clitoris and, 156n24

vaginal stimulation, 158n32, 159n3

van de Velde, Theodore, 156n22

van't Hof, Sophie R., 61, 68

Vasey, Paul L., 82, 83, 86, 88, 167n9

videos: erotic, 63, 161n21; pornographic, 63; sexual, 62

violence, 5, 166n3, 167n8; colonial, 10, 11, 100; genocidal, 10; racist, 100; sexual, 151n12

virginity pacts, 7

virility, 90–91

visual sexual stimuli (VSS), 62, 67, 162nn23–24; brain activation and, 63–64

voles, 100; compulsory sexuality and, 84–86; "monogamy" of, 84–86

voxels, 160n7

Walter, Martin, 67, 68, 165n32

Ward, Anna E., 63

Ward, Jane, 133

Warren, Wesley C., 118

Wehrum-Osinsky, Sina, 67–68, 161n18

Weismann, August, 113, 175n25

whiteness, 10, 62, 165n34

white settlers, 9, 11, 26–27, 151n13; "deviant" sexual interest and, 10

white supremacy, 15, 81, 151n13, 154n7

Wiens, John J., 114

Willey, Angela, 18, 84, 85, 100

Williams, George C., 110, 113, 176n25

Wilson, Elizabeth, 96

Wise, Nan: study by, 48, 49, 50–51, 159n1, 159n5

witchcraft, 27

Wolbachia, 122, 177n33

women's liberation movement, 6, 132

World Wildlife Fund, 76

Wright, Jessey, 53

Yerkes National Primate Research Center, 84

Young, Larry, 84

FEMINIST TECHNOSCIENCES
Rebecca Herzig and Banu Subramaniam, Series Editors

Figuring the Population Bomb: Gender and Demography in the Mid-Twentieth Century, by Carole R. McCann

Risky Bodies and Techno-Intimacy: Reflections on Sexuality, Media, Science, Finance, by Geeta Patel

Reinventing Hoodia: Peoples, Plants, and Patents in South Africa, by Laura A. Foster

Queer Feminist Science Studies: A Reader, edited by Cyd Cipolla, Kristina Gupta, David A. Rubin, and Angela Willey

Gender before Birth: Sex Selection in a Transnational Context, by Rajani Bhatia

Molecular Feminisms: Biology, Becomings, and Life in the Lab, by Deboleena Roy

Holy Science: The Biopolitics of Hindu Nationalism, by Banu Subramaniam

Bad Dog: Pit Bull Politics and Multispecies Justice, by Harlan Weaver

Underflows: Queer Trans Ecologies and River Justice, by Cleo Wölfle Hazard

Hacking the Underground: Disability, Infrastructure, and London's Public Transport System, by Raquel Velho

Queer Data Studies, edited by Patrick Keilty

Botany of Empire: Plant Worlds and the Scientific Legacies of Colonialism, by Banu Subramaniam

Acing Science: Compulsory Sexuality and Asexual Possibilities, by Kristina Gupta